B

Photo Bettina, Zürich

Complex Analysis

Articles dedicated
to Albert Pfluger
on the occasion of
his 80th birthday

Edited by
Joseph Hersch
and
Alfred Huber

1988   Birkhäuser Verlag
Basel · Boston · Berlin

Editors:
Prof. Joseph Hersch
Prof. Alfred Huber
ETH Zürich
Mathematik
ETH-Zentrum
CH-8092 Zürich

Library of Congress Cataloging in Publication Data
Complex Analysis: articles dedicated to Albert Pfluger on the
    occasion of his 80th birthday / edited by Joseph Hersch and Alfred
    Huber.
       p.   cm.
    Bibliographie: p.
    ISBN 3-7643-1958-5 : 98.00F (est.)
1. Functions of a complex variable.
2. Pfluger, Albert, 1907– . I. Pfluger,
Albert, 1907– . II. Hersch, Joseph, 1925– . III. Huber, Alfred, 1922– .
QA331.C6533 1988   515.9–dc19   88-19426 CIP

CIP-Kurztitelaufnahme der Deutschen Bibliothek

Complex analysis : articles dedicated to Albert Pfluger on the
occasion of his 80. birthday / ed. by Joseph Hersch and Alfred
Huber. – Basel ; Boston ; Berlin : Birkhäuser, 1988
    ISBN 3-7643-1958-5
NE: Hersch, Joseph [Hrsg.]; Pfluger, Albert: Festschrift

© 1988 Birkhäuser Verlag Basel
Typesetting and Layout: *mathScreen online*, CH-4056 Basel
Printed in Germany
ISBN 3-7643-1958-5
ISBN 0-8176-1958-5

Preface

The present volume contains articles pertaining to a wide variety of subjects such as conformal and quasiconformal mappings and related extremal problems, Riemann surfaces, meromorphic functions, subharmonic functions, approximation and interpolation, and other questions of complex analysis. These contributions by mathematicians from all over the world express consideration and friendship for Albert Pfluger. They reflect the wide range of his interests.

Albert Pfluger was born on 13 October 1907 in Oensingen (Kanton Solothurn) as the oldest son of a Swiss farmer. After a classical education he studied Mathematics at the ETH-Zurich. Among his teachers were Hopf, Plancherel, Pólya and Saxer. Pólya was his Ph.D. adviser. After some teaching at high schools (Gymnasien), he became professor at the University of Fribourg, and a few years later (1943) he was appointed as successor of Pólya at the ETH. He retired in 1978, but has always remained very active in research.

Pfluger's lectures were highly appreciated by the students. His vivid and clear teaching stimulated and challenged them to independent thinking. Many of his Ph.D. students are now themselves teaching in universities.

His main research relates to the following fields: entire functions, Riemann surfaces, quasiconformal mappings, schlicht functions. (See list of publications.) He collaborated with several mathematical colleagues, in particular with Rolf Nevanlinna, who taught parallel to him at the University of Zurich.

In 1973 Pfluger was nominated foreign member of the Finnish Academy of Sciences.

To Albert Pfluger, his wife Maria, their children and grandchildren we present our cordial wishes, and to the authors of this volume our sincere thanks.

Joseph Hersch, Alfred Huber

Publications of Albert Pfluger

[1] Über numerische Schranken im Schottky'schen Satz. Comment. Math. Helv., *7* (1934/35), 159–170.

[2] Über eine Interpretation gewisser Konvergenz- und Fortsetzungseigenschaften Dirichlet'scher Reihen. Comment. Math. Helv., *8* (1935/36), 3–43.

[3] On the power series of an integral function having an exceptional value (with G. Pólya). Proc. Cambridge Phil. Soc., *31* (pt. II) (1935), 153–155.

[4] Wachstum ganzer Funktionen. Verh. Schweiz. Naturforsch. Ges., Einsiedeln (1935), 280–281.

[5] On analytic functions bounded at the lattice points. Proc. London Math. Soc., Ser. 2, *42* (1937), 305–315.

[6] Über das Anwachsen von Funktionen, die in einem Winkelraum regulär und vom Exponentialtypus sind. Compositio Math., *4* (1937), 367–372.

[7] Sur la croissance et la distribution des zéros de certaines fonctions entières d'ordre positif fini. C.R. Acad. Sci. Paris, *205* (1937), 889–890.

[8] Sur la variation de l'argument et la distribution des zéros d'une certaine classe de fonctions analytiques. C.R. Acad. Sci. Paris, *206* (1938), 1786–1787.

[9] Die Wertverteilung und das Verhalten von Betrag und Argument einer speziellen Klasse analytischer Funktionen. Comment. Math. Helv., *11* (1938/39), 180–214 and *12* (1939/40), 25–65.

[10] Konforme Abbildung und eine Verallgemeinerung der Jensenschen Formel. Comment. Math. Helv., *13* (1940/41), 284–292.

[11] Über Interpolation ganzer Funktionen. Comment. Math. Helv., *14* (1941/42), 314–349.

[12] Über gewisse ganze Funktionen vom Exponentialtypus. Comment. Math. Helv., *16* (1943/44), 1–18.

[13] Über ganze Funktionen ganzer Ordnung. Comment. Math. Helv., *18* (1945/46), 177–203.

[14] Zur Defektrelation ganzer Funktionen endlicher Ordnung. Comment. Math. Helv., *19* (1946/47), 91–104.

[15] Sur l'unicité de la distribution de masses produisant un potentiel donné. Bull. Sci. Math. Paris, 2e sér., *71* (1947), 1–3.

[16] Bemerkungen zum Beitrag des Herrn P. Rossier über Funktionalgleichungen. Elemente der Math., *2* (1947), 12–13.

[17] Une propriété métrique de la représentation quasiconforme. C.R. Acad. Sci. Paris, *226* (1948), 623–625.

[18] Sur une propriété de l'application quasi conforme d'une surface de Riemann ouverte. C.R. Acad. Sci. Paris, *227* (1948), 25–26.

[19] La croissance des fonctions analytiques et uniformes sur une surface de Riemann ouverte. C.R. Acad. Sci. Paris, *229* (1949), 505–507.

[20] Des théorèmes du type de Phragmén-Lindelöf. C.R. Acad. Sci. Paris, *229* (1949), 542–543.

[21] Über das Anwachsen eindeutiger analytischer Funktionen auf offenen Riemann'schen Flächen. Ann. Acad. Sci. Fenn., Ser. A.I, *64* (1949), 1–18.

[22] Sur l'existence de fonctions non constantes, analytiques, uniformes et bornées sur une surface de Riemann ouverte. C.R. Acad. Sci. Paris, *230* (1950), 166–168.

[23] Quelques théorèmes sur une classe de fonctions pseudo-analytiques. C.R. Acad. Sci. Paris, *231* (1950), 1022–1023.

[24] A propos d'un mémoire récent de M. Brelot. Ann. Inst. Fourier, *2* (1950), 81–82.

[25] Quasikonforme Abbildungen und logarithmische Kapazität. Ann. Inst. Fourier, *2* (1950), 69–80.

[26] Généralisation du lemme de Schwarz et du principe de la mesure harmonique pour les fonctions pseudo-analytiques (with J. Hersch). C.R. Acad. Sci. Paris, *234* (1952), 43–45.

[27] Principe de l'augmentation des longueurs extrémales (with J. Hersch). C.R. Acad. Sci. Paris, *237* (1953), 1205–1207.

[28] Über das Typenproblem Riemann'scher Flächen. Comment. Math. Helv., *27* (1953), 346–356.

[29] Über die Riemann'sche Periodenrelation für offene Flächen. Proc. International Congress Math., Amsterdam (1954), vol. *1*, 458–459.

[30] Über die Bestimmung von obern und untern Schranken für die Kapazität und Torsionssteifigkeit. Verh. Schweiz. Naturforsch. Ges., Altdorf (1954), p. 98.

[31] Extremallängen und Kapazität. Comment. Math. Helv., *29* (1955), 120–131.

[32] Über die Riemannsche Periodenrelation auf transzendenten hyperelliptischen Flächen. Comment. Math. Helv., *30* (1956), 98–106.

[33] Ein Approximationssatz für harmonische Funktionen auf Riemannschen Flächen. Ann. Acad. Sci. Fenn., Ser. A.I, *216* (1956), 1–8.

[34] Ein alternierendes Verfahren auf Riemannschen Flächen. Comment. Math. Helv. *30* (1956), 265–274.

[35] Über ein simultanes Differenzenverfahren zur Abschätzung der Torsionssteifigkeit und der Kapazität nach beiden Seiten (with J. Hersch and A. Schopf). Zeitschr. Angew. Math. Phys. (ZAMP), *7* (1956), 89–113.

[36] Theorie der Riemannschen Flächen. Grundlehren der math. Wiss., *89* (Springer, 1957).

[37] A direct construction of Abelian differentials on Riemann surfaces. Semin. Analyt. Funct. (Inst. Adv. Study, Princeton), vol. *2* (1958), 39–48.

[38] Harmonische und analytische Differentiale auf Riemannschen Flächen. Ann. Acad. Sci. Fenn., Ser. A.I, *249/4* (1958), 1–18.

[39] Über die Äquivalenz der geometrischen und der analytischen Definition quasikonformer Abbildungen. Comment. Math. Helv., *33* (1959), 23–33.

[40] Über die Konstruktion Riemannscher Flächen durch Verheftung. J. Ind. Math. Soc., *24* (1960), 401–412.

[41] Verallgemeinerte Poisson-Stieltjes'sche Integraldarstellung und kontraktive Operatoren. Ann. Acad. Sci. Fenn., Ser. A.I, *336*/13 (1963), 1–14.

[42] Über harmonische Funktionen im Einheitskreis mit Werten in einem Banach'schen Raum.
(a) Verh. Schweiz. Naturforsch. Ges., Sitten (1963), 86–87.
(b) L'Enseignement Math., *10* (1964), 319–320.

[43] Zu einem Verzerrungssatz der konformen Abbildung. Math. Zeitschr., *84* (1964), 263–267.

[44] Verallgemeinerung eines Satzes von Pólya über den transfiniten Durchmesser ebener Punktmengen. Math. Zeitschr., *85* (1964), 285–290.

[45] Riemannsche Flächen vom hyperbolischen Typus, erzeugt durch Asymmetrien (with J. Sutter). Proc. Erevan Conference (1965), 253–257.

[46] On the convexity of some sections of the n-th coefficient body for schlicht functions.
(a) (in Russian) Some problems of mathematics and mechanics, volume dedicated to M.A. Lavrentjev for his 70th birthday, Acad. Sci. USSR, ed. "Nauka", Leningrad (1970), 233–241.
(b) Amer. Math. Soc. Transl. (2) *104* (1976), 215–222.

[47] Lineare Extremalprobleme bei schlichten Funktionen. Ann. Acad. Sci. Fenn., Ser. A.I, *489* (1971), 1–32.

[48] On a coefficient problem for schlicht functions. Springer Lecture Notes Math. *505* (Advances in complex function theory, Maryland, 1973/74), 79–91.

[49] Functions of bounded boundary rotation and convexity. J. d'Analyse Math., *30* (1976), 437–451.

[50] Some coefficient problems for starlike functions. Ann. Acad. Sci. Fenn., Ser. A.I, *2* (1976), 383–396.

[51] On a uniqueness theorem in conformal mapping. Michigan Math. J., *23* (1976), 363–365.

[52] On a coefficient inequality for schlicht functions. Springer Lecture Notes Math. *743* (Romanian-Finnish Seminar on Complex Analysis, Bucharest, 1976), 336–343.

[53] George Pólya. J. Graph Theory, *1* (1977), 291–294.

[54] Über die Koeffizienten schlichter Funktionen. Bonner Math. Schriften, *121* (1980), 41–61.

[55] Non-linear extremal problems for starlike functions. J. d'Analyse Math., *36* (1979), 217–226.

[56] On the diameter of planar curves and Fourier coefficients.
(a) J. Appl. Math. Phys. (ZAMP), *30* (1979), 305–314.
(b) Colloq. Math. Soc. János Bolyai, (35) (Budapest, 1980), 957–965.

[57] Die Bedeutung der Arbeiten Christoffels für die Funktionentheorie. E.B. Christoffel, ed. by P.L. Butzer and F. Fehér (Birkhäuser, 1981), 244–252.

[58] Some extremal problems for functions of bounded boundary rotation (with R. Boutellier). Israel J. of Math., *39* (1981), 46–62.

[59] On support points in the class of functions with bounded boundary rotation (with W. Hengartner and G. Schober). Ann. Acad. Sci. Fenn., Ser. A.I, *6* (1981), 213–224.

[60] Über konforme Abbildungen des Einheitskreises. Ann. Acad. Sci. Fenn., Ser. A.I, *7* (1982), 73–79.

[61] Bemerkung zum Beitrag von R. Mortini. Elemente der Math., *38* (1983), 102–103.

[62] Über eine die konvexen Kurven kennzeichnende Minimaleigenschaft. Elemente der Math., *38* (1983), 113–119.

[63] Close-to-convex functions and functions of bounded boundary rotation. Complex Variables, *3* (1984), 205–210.

[64] Varianten des Schwarzschen Lemma. Elemente der Math., *40* (1985), 46–47.

[65] The Fekete-Szegö inequality by a variational method. Ann. Acad. Sci. Fenn., Ser. A.I, *10* (1985), 447–454.

[66] The Fekete-Szegö inequality for complex parameters. Complex Variables, *7* (1986), 149–160.

[67] The second coefficient body of $\sum$. J. d'Analyse Math., *46* (1986), 221–229.

[68] On a method of Georg Faber. Complex Variables, *9* (1987), 251–261.

[69] On the functional $a_3 - \lambda a_2^2$ in the class S. Complex Variables, *10* (1988), 83–95.

Table of Contents

| Complex
Analysis | Edited by
J. Hersch and A. Huber | Birkhäuser Verlag
Basel 1988 |

Lars V. Ahlfors

Cross-ratios and Schwarzian Derivatives in R^n

This paper was written several years ago, but no part of it has been published previously. A preprint was distributed to selected experts and seems to have been favorably received. For some time I had hoped to improve on the results of the paper, but as years went by my research took a different direction, and it became implausible that I would add anything significant to the paper as it stands.

Meanwhile there has been considerable progress in this area, but my friends have insisted that the bulk of the paper still has at least some historical interest and should be made available to the mathematical public. It gives me great satisfaction that the paper will appear in its original form in this volume dedicated to Professor Albert Pfluger in appreciation of his lasting contributions to analysis

The research for the paper was supported by the National Science Foundation.

The cross-ratio is of fundamental importance in projective geometry and some aspects of complex function theory. In the latter connection the cross-ratio of four complex numbers a, b, c, d is defined as

$$(a, b, c, d) = \frac{a - c}{a - d} : \frac{b - c}{b - d},$$

another complex number. On the other hand, in geometry the cross-ratio occurs mainly as a double ratio $(AC/AD) : (BC)/(BD)$ of the lengths of four segments.

Recent developments in function theory, especially in connection with Kleinian groups, have made it even more essential than at the time of Poincaré and Klein to study the conformal structure of three-space as an extension of the conformal structure of the complex plane. Experience has shown that many methods which carry over effortlessly from two to three dimensions do not extend to arbitrary R^n. For this reason it seems to the author that the case of arbitrary dimension is not an idle generalization, but may serve to throw new light on the cases $n = 2$ and 3 as well. With some degree of justification it can be maintained that a method which does not generalize is not fully understood.

The theory of Möbius transformations does of course generalize almost automatically, but the lack of a natural generalization of the complex cross-ratio has been a considerable handicap. One of the purposes of this paper is to suggest a way to overcome this difficulty. Since the idea is quite simple it may have occured to others as well, but since I am not aware of any mention of it in the literature I have thought it worth while to give it some publicity.

The second part of the paper is devoted to a study of the Schwarzian derivative. It is commonly accepted that the Schwarzian derivative is an infinitesimal version of the cross-ratio, but this is seldom made explicit. Traditionally, the Schwarzian is considered only in connection with holomorphic functions of one variable, and in that case the relation to the cross-ratio is fairly obvious. It seems that the more general case of smooth mappings into R^n has hardly been explored at all. It turns out that the real and imaginary parts of the Schwarzian can both be generalized, albeit in somewhat different ways. At present these generalizations are more or less tentative, and there are no significant applications, but in view of the importance of the holomorphic Schwarzian it is not unreasonable to make at least a preliminary forage in this direction.

§1 Möbius transformations and cross-ratios.

1. When A.F. Möbius introduced the notion of what we call a Möbius transformation he did not connect it with the idea of a fractional linear transformation with complex coefficients, nor did he regard conformality as the main feature. His was a purely geometric theory of "Kreisverwandtschaften", a term that defies translation. In modern terminology a "Kreisverwandtschaft" is a homeomorphism of the extended complex plane $\bar{C}$ which maps circles on circles. This led him to the invariance of the cross-ratio and the angle, but he made only minimal use of the complex notation. It is interesting to note that Möbius was well aware that his definition works equally well in three dimensions. If it had not been considered esoteric at the time he would probably have used n dimensions.

2. The modern approach is much more direct. We begin with the complex case and define a Möbius transformation by the formula

$$\gamma z = \frac{az + b}{cz + d} \tag{1}$$

where $a, b, c, d \in C$ and $ad - bc = 1$. We regard (1) as a mapping of $\bar{C}$ on itself with the standard conventions for ∞.

From (1) one derives the difference formula

$$\gamma z - \gamma\zeta = \frac{z - \zeta}{(cz + d)(c\zeta + d)} \tag{2}$$

which implies the existence of the derivative

$$\gamma'(z) = (cz + d)^{-2}. \tag{3}$$

We rewrite (2) as

$$\gamma z - \gamma\zeta = \gamma'(z)^{1/2}\gamma'(\zeta)^{1/2}(z - \zeta). \tag{4}$$

We shall find (4) an extremely useful tool even, within certain limits, in the multidimensional case. For the moment we observe merely that it proves the invariance of the cross-ratio. Indeed, with the definition

$$(z, z', \zeta, \zeta') = (z - \zeta)(z - \zeta')^{-1}(z' - \zeta')(z' - \zeta)^{-1}$$

it follows that $(\gamma z, \gamma z', \gamma\zeta, \gamma\zeta') = (z, z', \zeta, \zeta')$, for the factors $\gamma'(z)^{1/2}$ etc. introduced by (4) cancel against each other.

We remark in passing that the cross-ratio is well defined, finite or infinite, as soon as no more than two of the z, z', ζ, ζ' are equal.

3. It is useful to recall that four numbers determine six cross-ratios, depending on the order. Each corresponds to four permutations: in simplified notation $(abcd) = (cdab) = (dabc) = (dcba)$.

There is a unique Möbius tranformation that carries three given distinct points b, c, d into $0, 1, \infty$ in this order. Therefore, there is a unique complex number z such that $(abcd) = (z, 1, 0, \infty) = z$, $(acbd) = (z, 0, 1, \infty) = 1 - z$, $(bacd) = z^{-1}$. The other three cross-ratios are $(1 - z)^{-1}$, $1 - z^{-1}$, $(1 - z^{-1})^{-1}$. It is sufficient, however, to retain the basic relations

$$(abcd) + (acbd) = 1, \quad (abcd)(bacd) = 1. \tag{5}$$

They remain in force even in the event of one or two pairs of equal numbers.

In addition to $(z, 1, 0, \infty)$ there is another normal form for the cross-ratio, namely $(-1, -e^\tau, e^\tau, 1)$. The condition $z = (-1, -e^\tau, e^\tau, 1)$ translates to $z = ch^2(\tau/2)$. This means that τ is determined up to sign and additive multiples of $2\pi i$. It becomes unique if we require that $0 \leq \operatorname{Im}\tau < \pi, \tau = 0$ if τ is real. In relation to $z = (abcd)$ the number τ, so normalized, is referred to as the *complex distance* between the ordered pairs (a, d) and (b, c). Its geometric meaning will be explained later. It seems to have been first introduced by F. Schilling in 1891.

4. We pass now to R^n, the complex plane being identified with R^2. We shall use the notations $x = (x_1, \ldots, x_n) \in R^n$, $|x|^2 = x_1^2 + \ldots + x_n^2$, and $(x, y) = x_1 y_1 + \ldots + x_n y_n$. As usual, R^n is compactified to $\bar{R}^n = R^n \cup \{\infty\}$. A *similarity* is a mapping $\bar{R}^n \to \bar{R}^n$ whose restriction to R^n is given by $x \to mx + b$, where $b \in R^n$ and m is a *conformal matrix*, i.e. a matrix λk, $\lambda > 0$, $k \in O(n)$. Also, ∞ is mapped on itself.

The *inversion*, or reflection in the unit sphere S^{n-1}, is defined by $x \to x^* = x/|x|^2$ when $x \neq 0, \infty$ and $0^* = \infty$, $\infty^* = 0$.

Definition 1. The group $M(\bar{R}^n)$ of Möbius transformations is the group generated by all similarities together with the inversion in the unit sphere.

If $\Omega \subset R^n$ is open the derivative of a mapping $f : \Omega \to R^n$ at $x \in \Omega$, if it exists, is the matrix $f'(x)$ or $Df(x)$ with elements $f'(x)_{ij} = \partial f_i / \partial x_j$. Clearly, $D(mx+b) = m$ and, by elementary calculation, $Dx^* = |x|^{-2}(\delta_{ij} - 2x_i x_j/|x|^2)$. In this paper we shall use the notation $Q(x)$ for the matrix with elements $Q(x)_{ij} = x_i x_j/|x|^2$ and I or I_n for the unit matrix. With this notation

$$Dx^* = |x|^{-2}(I - 2Q(x)). \tag{6}$$

One verifies that $Q(x)^2 = Q(x)$ and $(I - 2Q(x))^2 = I$, $I - 2Q(x) \in O(n)$. Matrices of the form $I - 2Q(a)$ will occur frequently. They have a simple geometric interpretation: $(I - 2Q(a))x$ is the mirror image of x with respect to the hyperplane through 0 perpendicular to a.

According to (6) Dx^* is a conformal matrix, and by the chain rule the derivative $\gamma'(x)$ of any $\gamma \in M(\bar{R}^n)$ is likewise a conformal matrix. In other words, the mapping by a Möbius transformation is conformal. For $n > 2$ the converse is a classical theorem due to Liouville.

As a conformal matrix $\gamma'(x)$ can be written in the form λk with $\lambda > 0$, $k \in O(n)$; unless γ is a similarity λ and k will depend on x. We shall denote λ by $|\gamma'(x)|$; because of the conformality λ is also the operator norm of the matrix $\gamma'(x)$ and the linear change of scale at x, the same in all directions.

The determinant of $k = \gamma'(x)/|\gamma'(x)|$ is constantly 1 if γ is sense-preserving, -1 if it is sense-reversing. It is possible to restrict attention to the sense-preserving subgroup, but this is not always an advantage. We prefer to stay with the original definition of $M(\bar{R}^n)$ as the group of all Möbius transformations.

As customary, we shall identify R^{n-1} with the set of $x \in R^n$ with $x_n = 0$. The points with $x_n > 0$ form the upper half-space H^n. We denote by $M(H^n)$ the subgroup which maps H^n on itself. Similarly, $M(B^n)$ will be the subgroup that preserves the unit ball. The groups $M(H^n)$, $M(B^n)$ and $M(R^{n-1})$ are isomorphic.

5. Formula (4), restricted to absolute values, remains valid in R^n.

Proposition 1. If $\gamma \in M(\bar{R}^n)$, then

$$|\gamma x - \gamma y| = |\gamma'(x)|^{1/2} |\gamma'(y)|^{1/2} |x - y| \tag{7}$$

for all $x, y \in R^n \setminus \gamma^{-1}\infty$.

The formula is trivial when γ is a similarity. For $\gamma x = x^*$ it reduces to $|x^* - y^*| = |x|^{-1}|y|^{-1}|x - y|$ which is easily verified. The general validity of (7) follows by the chain rule.

If $x, y, u, v \in R^n$ there is no immediate way of forming a cross-ratio since multiplication has no meaning. However, if we use only distances we can still form the *absolute cross-ratio*

$$|x, u, v, y| = |x - v|\,|x - y|^{-1}|u - y|\,|u - v|^{-1}. \tag{8}$$

It is again well defined as long as no three points coincide and we admit ∞ as a possible value.

Proposition 2. $|\gamma x, \gamma u, \gamma v, \gamma y| = |x, u, v, y|$ for every $\gamma \in M(\bar{R}^n)$.

This is a trivial consequence of (7).

6. We shall use Proposition 2 to prove:

Proposition 3. Every $\gamma \in M(\bar{R}^n)$ with $\gamma\infty = \infty$ is a similarity.

Since $\gamma - \gamma 0$ has fixed points at 0 and ∞ we may as well assume that $\gamma 0 = 0$, $\gamma\infty = \infty$ and show that $\gamma x = mx$ with a constant conformal matrix m. By Proposition 2, $|\gamma x, \gamma y, 0, \infty| = |x, y, 0, \infty|$ or $|\gamma x|/|\gamma y| = |x|/|y|$, and similarly $|\gamma x - \gamma y|/|\gamma y| = |x - y|/|y|$. From the first relation $|\gamma x| = \lambda|x|$ with constant λ. From the second $|\gamma x - \gamma y|^2 = \lambda^2|x - y|^2$ and hence $(\gamma x, \gamma y) = \lambda^2(x, y)$. On expanding the squares it follows that $|\gamma(x + y) - \gamma x - \gamma y|^2 = \lambda^2|(x + y) - x - y|^2 = 0$ and thus $\gamma(x + y) = \gamma x + \gamma y$, $\gamma'(x + y) = \gamma'(x)$, a constant.

7. Proposition 3 leads to a simple normal form for all Möbius transformations. For given γ we shall write $\gamma^{-1}0 = u$, $\gamma^{-1}\infty = v$ and assume that $v \neq \infty$. Then $\sigma x = (x - v)^* - (u - v)^*$ is a Möbius transformation with $\sigma u = 0$, $\sigma v = \infty$ so that $\sigma\gamma^{-1}$ has 0 and ∞ as fixed points. We conclude by Proposition 3 that

$$\gamma x = m[(x - v)^* - (u - v)^*] \tag{9}$$

where m is a constant conformal matrix.

Sometimes it is preferable to replace (9) by

$$\gamma x = m[(x^* - v^*)^* - (u^* - v^*)^*] \tag{10}$$

provided that u and v are different from 0. Since every mapping of the form (9) is also of the form (10) there exists a conformal m such that

$$(x^* - v^*)^* - (u^* - v^*)^* = m[(x - v)^* - (u - v)^*]. \tag{11}$$

To find m we shall first compare the absolute values. By (6) and (7)

$$|(x^* - v^*)^* - (u^* - v^*)^*| = |x^* - u^*|/|x^* - v^*|\,|u^* - v^*|$$
$$= |x - u|\,|v|^2/|x - v|\,|u - v|$$

and $|(x - v)^* - (u - v)^*| = |x - u|/|x - v|\,|u - v|$ so that $m = |v|^2 k$, $k \in O(n)$. To determine k we differentiate (11). By (6) and the chain rule we obtain

$$(I - 2Q(x^* - v^*))\,(I - 2Q(x)) = k(I - 2Q(x - v)). \tag{12}$$

For $x = 2v$ the matrices $I - 2Q$ in this formula are all equal to $I - 2Q(v)$ so that (12) gives $k = I - 2Q(v)$. At the same time we have proved the identity $(I - 2Q(x^* - v^*))\,(I - 2Q(x)) = (I - 2Q(v))\,(I - 2Q(x - v))$ or, in different notation

$$I - 2Q(a^* - b^*) = (I - 2Q(a))\,(I - 2Q(a - b))\,(I - 2Q(b)). \tag{13}$$

We shall choose

$$\gamma_{uv}x = (x^* - v^*)^* - (u^* - v^*)^* \tag{14}$$

to be the standard mapping with $\gamma u = 0$, $\gamma v = \infty$. It is useful to display the alternative expression

$$\gamma_{uv}x = |v|^2(I - 2Q(v))\,((x - v)^* - (u - v)^*) \tag{15}$$

as well as the formulas

$$\gamma'_{uv}(x) = |v|^2 |x - v|^{-2} (I - 2Q(v))\,(I - 2Q(x - v)) \qquad (16)$$

and

$$|\gamma_{uv}x| = \frac{|v|^2 |x - u|}{|u - v|\,|x - v|}. \qquad (17)$$

The case $|u| < 1$, $v = u^*$ is particularly important. In earlier papers ([1]–[4]) I have denoted the mapping

$$(1 - |u|^2)\gamma_{uu^*}x = -u + (1 - |u|^2)\,(x^* - u)^* \qquad (18)$$

by T_u. It is a standard mapping of the unit ball on itself with u going to 0.

Proposition 4. If $\gamma \in M(\bar{R}^n)$, then

$$I - 2Q(\gamma x - \gamma y) = (\gamma'(x)/|\gamma'(x)|)^{-1}(I - 2Q(x - y))\,(\gamma'(y)/|\gamma'(y)|) \qquad (19)$$

for all $x, y \in R^n$, $x \neq y$.

This is a counterpart of Proposition 1. The formula is trivial when γ is a similarity. For $\gamma x = x^*$ it is proved by (13), and in the general case it follows by the chain rule. Observe that x and y are interchangeable.

8. We recall that $I - 2Q(a)$ represents reflection in a hyperplane. A product $(I - 2Q(a))\,(I - 2Q(b))$ is thus a rotation by an angle which is twice the angle formed by a and b. This motivates us to introduce, in analogy with (8), an *angular cross-ratio*, defined by

$$\Phi(x, y, u, v)$$
$$= (I - 2Q(x - u))\,(I - 2Q(x - v))\,(I - 2Q(y - v))\,(I - 2Q(y - u)). \qquad (20)$$

The absence of inverses is due to the relation $(I - 2Q)^2 = I$.

Proposition 5. The conjugacy class and hence the trace of $\Phi(x, y, u, v)$ are invariant under the Möbius group $M(\bar{R}^n)$. More precisely,

$$\Phi(\gamma x, \gamma y, \gamma u, \gamma v) = (I - 2Q(u))\Phi(x, y, u, v)\,(I - 2Q(u)). \qquad (21)$$

This is a direct consequence of (19).

9. We turn to the problem of defining a complex cross-ratio (x, y, u, v) of four points in R^n. In principle this is very easy. Any four points in R^n lie on a two-sphere (which may degenerate to a plane). The two-sphere can be mapped by a similarity on the unit sphere in R^3, spanned by the first three coordinate vectors e_1, e_2, e_3. We identify e_1, e_2 with $1, i$ and map the sphere on the complex plane by stereographic projection. The points are now represented by complex numbers x', y', u', v' and we define $(x, y, u, v) = (x', y', u', v')$, a complex number. In the degenerate case the stereographic projection is not needed.

There is a slight catch. The mapping of x, y, u, v on x', y', u', v' can be either sense-preserving or sense-reversing. For this reason (x', y', u', v') is unique only up to complex conjugation. To achieve uniqueness it is necessary to make an arbitrary choice, and we shall do it by requiring the imaginary part of (x, y, u, v) to be $= 0$. We adopt the following definition:

Definition 2. The complex cross-ratio (x, y, u, v) shall be the unique complex number $z = \xi + i\eta$ with $\eta = 0$ for which there exists a $\gamma \in M(\bar{R}^n)$ which maps x, y, u, v on $\xi e_1 + \eta e_2, e_1, 0, \infty$.

We leave it to the reader to prove the uniqueness. Although the complex cross-ratio has not been defined as a ratio of differences it retains the properties of the ordinary cross-ratio with minor modifications. The invariance under Möbius transformations is immediate from the definition. The cross-ratios $z = (x, y, u, v)$ and $z'(x, u, y, v)$ are related by $z + \bar{z}' = 1$, $z\bar{z}' = 1$. The easiest way to find z is from $|z| = |x, y, u, v|$, $|z - 1| = |x, u, y, v|$.

10. The next step is to clarify the relation between the complex, the absolute, and the angular cross-ratios. We begin by determining $I - 2Q(a)$ for $n = 2$ with the complex notation $a = a_1 + ia_2$. One finds at once

$$I - 2Q(a) = \begin{pmatrix} -\operatorname{Re}(a/\bar{a}) & -\operatorname{Im}(a/\bar{a}) \\ -\operatorname{Im}(a/\bar{a}) & \operatorname{Re}(a/\bar{a}) \end{pmatrix} \tag{22}$$

Complex numbers $c = c_1 + ic_2$ correspond isomorphically to matrices $\begin{pmatrix} c_1 & c_2 \\ -c_2 & c_1 \end{pmatrix}$. Multiplication with $j = \begin{pmatrix} -1 & 0 \\ 0 & 1 \end{pmatrix}$ yields

$$jc = \bar{c}j = \begin{pmatrix} -c_1 & -c_2 \\ -c_2 & c_1 \end{pmatrix}.$$

Since $j^2 = I$ it follows that $\Phi(x, y, u, v)$ can be identified with the complex number

$$\frac{(\bar{x} - \bar{u})(x - v)(y - v)(\bar{y} - \bar{v})}{(x - u)(\bar{x} - \bar{v})(\bar{y} - \bar{v})(y - u)} = (x, y, u, v)^-/(x, y, u, v).$$

This is not immediately applicable to $z = (z, 1, 0, \infty)$. However, we remarked in **3** that $z = (z, 1, 0, \infty) = (-1, -e^\tau, e^\tau, 1) = ch^2(\tau/2)$. The definition of Φ applies to the second form of the cross-ratio and we conclude that

$$\Phi(x, y, u, v) = \bar{z}/z = (x, y, u, v)^-/(x, y, u, v) \qquad (23)$$

in the sense of the standard identification of complex numbers with 2×2 matrices.

With minor changes this identification remains valid for $n > 2$ and arbitrary $x, y, u, v \in R^n$. If the complex cross-ratio is $\rho e^{i\theta}$, then $|x, y, u, v| = \rho$ while $\Phi(x, y, u, v)$ is conjugate to the matrix $\begin{pmatrix} \cos 2\theta & -\sin 2\theta \\ \sin 2\theta & \cos 2\theta \end{pmatrix}$ extended by ones in the diagonal. In particular, the trace of Φ is $n - 2 + 2\cos 2\theta$.

The geometric significance of θ is clear. Since θ is the angle between the rays from 0 to ∞ through 1 and z it is also, by conformality, the angle between the circular arcs in R^n passing from u through x to v and from u through y to v. These are directed arcs, but the angle is always between 0 and π, inclusive, and hence insensitive to an interchange of x and y.

11. This chapter would not be complete without a discussion of the upper half-space H^{n+1} and its hyperbolic geometry. The space H^{n+1} consists of all $x = (x_1, \ldots, x_{n+1})$ with $x_{n+1} > 0$; we use $\bar{x} = (x_1, \ldots, -x_{n+1})$ for the symmetric point in the lower half-space. The Möbius transformations in $\bar{R}^n$ extend automatically to transformations in $\bar{R}^{n+1}$ characterized by $(\gamma x)_{n+1} > 0$ when $x_{n+1} > 0$ and $\gamma \bar{x} = \overline{\gamma x}$. The extensions form the group $M(H^{n+1})$.

The invariance of $|x, \bar{x}, y, \bar{y}| = |x - y|^2/|x - \bar{y}|^2$ shows that the pseudo-distance $\delta(x, y) = |x - y|/|x - \bar{y}|$ is invariant, and it follows that $ds = |dx|/x_{n+1}$ is an invariant metric, the Poincaré metric for the half-space (the corresponding metric in the unit ball is $ds = 2|dx|/(1 - |x|^2)$). The geodesics of this metric are the vertical half-lines and the semicircles orthogonal to R^n. They are straight lines of the hyperbolic geometry of the half-space.

If x and y lie on the geodesic from 0 to ∞ their hyperbolic distance $d(x, y)$ is $\log |y|/|x|$, or rather its absolute value; it is more appropriate, however, to consider the geodesic as a directed line from 0 to ∞ and $\log |y|/|x|$ as a signed distance, positive if x and y follow each other in the direction of the line. It can also be expressed as $\log |y, x, 0, \infty|$.

An arbitrary geodesic is determined by its end points $u, v \in \bar{R}^n$ in this order, the geodesic being denoted by (u, v). Since (u, v) can be mapped on $(0, \infty)$ by a Möbius transformation the directed distance between two points x and y on (u, v) is measured by $\log |x, y, u, v|$. Another convenient formula for $d = d(x, y)$, counted positive, is $\tanh \frac{d}{2} = |x - y|/|x - \bar{y}|$.

The shortest distance from a point $x \in H^{n+1}$ to $(0, \infty)$ is along the geodesic through x which meets $(0, \infty)$. If the angle between x and e_{n+1} is ϕ it follows by the formula quoted above that the distance d between the point and the line is given by $\tanh \frac{d}{2} = \tan \frac{\phi}{2}$. This is equivalent to $\sinh d = \tan \phi$, $\tanh d = \sin \phi$, $\cosh d = \sec \phi$. To obtain an invariant formula we observe that $|0, x, \bar{x}, \infty| = \frac{1}{2} \sec \phi$ and thus $\cosh d = 2|0, x, \bar{x}, \infty|$. It follows that the distance from x to an arbitrary geodesic (u, v) is determined by $\cosh d = 2|u, x, \bar{x}, v|$.

Any two non-intersecting geodesics (x, y) and (u, v) have a unique common normal. Indeed, since the end points are at infinite distance the existence of two points with minimal distance is clear by compactness, and the geodesic through these points is orthogonal to the given geodesics.

In order to find the distance between the given geodesics we assume first that their common normal is $(0, \infty)$. In that case $y = -x$ and $v = -u$, and the intersections with the common normal are at $|x|e_{n+1}$ and $|u|e_{n+1}$. We conclude that the shortest distance is $d = \log(|x|/|u|)$, up to sign. On the other hand $|-x, -u, u, x| = |x + u|^2/4|x|\,|u|$ and $|-x, u, -u, x| = |x - u|^2/4|x|\,|u|$. It follows that $|-x, -u, u, x| + |-x, u, -u, x| = (|x|^2 + |u|^2)/2|x|\,|u| = \cosh d$. By invariance, the distance between (x, y) and (u, v) is hence given by

$$\cosh d = |y, v, u, x| + |y, u, v, x|. \tag{24}$$

We recall that the complex cross-ratios satisfy $(y, v, u, x) + (y, u, v, x)^- = 1$. Together with (24) it follows that $d = 0$ if and only if $0 < (y, v, u, x) < 1$. In other words, the geodesics intersect if and only if the points x, y, u, v lie on a circle with x, y separating u, v.

12. We return to the normalization $(-x, -u, u, x)$ with $(0, \infty)$ the common normal of $(x, -x)$ and $(u, -u)$. On passing to the complex cross-ratio we can replace $(-x, -u, u, x)$ by $(-1, -e^\tau, e^\tau, 1)$. One has $(-1, -e^\tau, e^\tau, 1) = \cosh^2 \frac{\tau}{2}$ and $(-1, e^\tau, -e^\tau, 1) = -\sinh^2 \frac{\tau}{2}$. Therefore, the distance d between $(x, -x)$ and $(u, -u)$ satisfies $\cosh d = |\cosh^2 \frac{\tau}{2}| + |\sinh^2 \frac{\tau}{2}| = \cosh \frac{\tau}{2} \cosh \frac{\bar{\tau}}{2} + \sinh \frac{\tau}{2} \sinh \frac{\bar{\tau}}{2} = \cosh \frac{\tau + \bar{\tau}}{2}$. We conclude that the positive distance is $d = |\mathrm{Re}\,\tau|$.

There is also a simple interpretation of $\mathrm{Im}\,\tau$. The imaginary part of τ is nothing else than the angle between the segments $(-1, 1)$ and $(-e^\tau, e^\tau)$. It is also the angle between the tangents to the two geodesics at their point of intersection with the common normal. This angle has invariant meaning if interpreted as obtained by parallel displacement along the normal. As in **3** we choose τ so that $0 \leq \mathrm{Im}\,\tau < \pi$, and we refer to τ as the complex distance between the directed geodesics $(-1, 1)$ and $(-e^\tau, e^\tau)$. In the general situation τ is the complex distance between (x, y) and (u, v) if $(x, u, v, y) = (-1, -e^\tau, e^\tau, 1)$ up to conjugation.

Remark. The term complex distance was used by F. Schilling in 1891, but it may be older.

§2 Möbius transformations and Schwarzian derivatives

1. We recall that the Schwarzian derivative of a function f is defined by

$$Sf = f'''/f' - \frac{3}{2}(f''/f')^2 = (f''/f')' - \frac{1}{2}(f''/f')^2. \tag{1}$$

There are two cases: either $f(z)$ is an analytic function of a complex variable, or $f(t)$ is a C^3 function of a real variable.

The generalization to vector valued functions must avoid using the quotients f''/f' and f'''/f' which have no meaning. This is easy as far as the real part of the Schwarzian is concerned, for we have clearly

$$\operatorname{Re} Sf = (f''', f')/|f'|^2 - 3(f'', f')^2/|f'|^4 + \frac{3}{2}|f''|^2/|f'|^2. \tag{2}$$

The expression on the right makes sense in any dimension, and we shall denote it by $S_1 f$.

In order to find a similar expression for the imaginary part we begin by writing, in the complex case,

$$Sf = [|f'|^2 f''' \bar{f}' - \frac{3}{2}(f'' \bar{f}')^2]/|f'|^4 \tag{3}$$

from which we obtain

$$\operatorname{Im} Sf = [|f'|^2 \operatorname{Im}(f''' \bar{f}') - 3(f', f'')\operatorname{Im}(f'' \bar{f}')]/|f'|^4. \tag{4}$$

To interpret (4) in the general case we shall think of f', f'', f''' as vectors in R^n, and we introduce the wedge notation $f' \wedge f''$ for the bivector with components $(f' \wedge f'')_{ij} = f'_i f''_j - f'_j f''_i$. The square norm is defined by

$$|f' \wedge f''|^2 = \sum_{i<j}(f' \wedge f'')_{ij}^2 \tag{5}$$

which can also be written as

$$|f' \wedge f''|^2 = |f'|^2|f''|^2 - (f', f'')^2. \tag{6}$$

In the complex case this means that $(\operatorname{Im} f'' \bar{f}')^2 = |f' \wedge f''|^2$ and $(\operatorname{Im} f''' \bar{f}')^2 = |f' \wedge f'''|^2$.

In generalization of (6)

$$(f' \wedge f'', f' \wedge f''') = -(f', f'')(f', f''') + |f'|^2(f'', f''')$$
$$= (\operatorname{Im} f'' \bar{f}')\operatorname{Im}(f''' \bar{f}'). \tag{7}$$

We now introduce the notation $S_2 f$ for the bivector

$$S_2 f = (f' \wedge f''')/|f'|^2 - 3(f', f'')(f' \wedge f'')/|f'|^4. \tag{8}$$

An easy computation that makes use of (4), (6) and (7) shows that, in the complex case, $(\operatorname{Im} Sf)^2 = |S_2 f|^2$. It is therefore not unreasonable to regard $S_2 f$ as a generalization of $\operatorname{Im} Sf$.

2. We continue with the complex case. In order to present the Schwarzian $Sf(z)$ as a limiting case of the cross-ratio we pick four distinct complex numbers a, b, c, d and develop the cross-ratio

$$Sf(z,t) = (f(z+ta), f(z+tb), f(z+tc), f(z+td))$$

in powers of t.

From the Taylor development

$$f(z+ta) = f(z) + af'(z)t + \frac{1}{2}a^2 f''(z)t^2 + \frac{1}{6}a^3 f'''(z)t^3 + \ldots$$

we obtain, for instance,

$$f(z+ta) - f(z+tc)$$
$$= (a-c)f't\left[1 + \frac{1}{2}(a+c)\frac{f''}{f'}t + \frac{1}{6}(a^2+ac+c^2)\frac{f'''}{f'}t^2 + \ldots\right]$$

and

$$\log(f(z+ta) - f(z+tc)) = \log[(a-c)f't] + \frac{1}{2}(a+c)\frac{f''}{f'}t +$$
$$+ \left[\frac{1}{6}(a^2+ac+c^2)\frac{f'''}{f'} - \frac{1}{4}(a+c)^2\left(\frac{f''}{f'}\right)^2\right]t^2 + \ldots \tag{9}$$

In view of $(a+c) - (a+d) + (b+d) - (b+c) = 0$ and

$$(a^2+ac+c^2) - (a^2+ad+d^2) + (b^2+bd+d^2) - (b^2+bc+c^2)$$
$$= \frac{1}{2}[(a+c)^2 - (a+d)^2 + (b+d)^2 - (b+c)^2] = (a-b)(c-d)$$

it follows easily that

$$\log Sf(z,t) = \log(a,b,c,d) + \frac{1}{6}(a-b)(c-d)Sf(z)t^2 + \ldots \qquad (10)$$

and thus

$$Sf(z,t) = (a,b,c,d)\left(1 + \frac{1}{6}(a-b)(c-d)Sf(z)t^2 + o(t^2)\right). \qquad (11)$$

This exhibits the asymptotic relationship between $Sf(z,t)$ and $Sf(z)$.

3. We pass to the case of a sufficiently smooth mapping $f : R^n \to R^n$. The generalized real and imaginary parts of Sf given by (2) and (8) will henceforth be denoted by $S_1 f$ and $S_2 f$ respectively. We shall find that $S_1 f$ has an interpretation similar to (11) while $S_2 f$ is best interpreted in the setting of differential geometry of space curves. Because the computation is not very different from the preceding one we shall present only the salient features.

The derivatives of f will be taken at the point x in a fixed direction u. In other words, they are homogeneous polynomials of the u_i defined by the expansion

$$f(x+tu) = f(x) + f'(x)t + \frac{1}{2}f''(x)t^2 + \frac{1}{6}f'''(x)t^3 \ldots .$$

This time a,b,c,d will be *real* numbers, and we investigate the absolute cross-ratio $|Sf(x,t)| = |f(x+tau), f(x+tbu), f(x+tcu), f(x+tdu)|$. One finds

$$|f(x+atu) - f(x+ctu)|^2 = |a-c|^2|f'|^2 t^2 \left\{1 + (a+c)\frac{(f',f'')}{|f'|^2}t + \right.$$

$$\left. + \left[\frac{1}{3}(a^2+ac+c^2)\frac{(f',f''')}{|f'|^2} + \frac{1}{4}(a+c)^2\frac{|f''|^2}{|f'|^2}\right]t^2 + \ldots\right\}$$

and

$$\log|f(x+atu) - f(x+ctu)| = \log(|a-c||f'|t) + \frac{1}{2}(a+c)\frac{(f',f'')}{|f'|^2}t +$$

$$+ \left[\frac{1}{6}(a^2+ac+c^2)\frac{(f',f''')}{|f'|^2} + \frac{1}{8}(a+c)^2\left(\frac{|f''|^2}{|f'|^2} - 2\frac{(f',f'')^2}{|f'|^4}\right)\right]t^2 + \ldots .$$

Exactly as in the step from (9) to (10) this leads to

$$\log|Sf(x,t)| = \log|a,b,c,d| +$$

$$+ \frac{1}{6}(a-b)(c-d)\left[\frac{(f',f''')}{|f'|^2} + \frac{3}{2}\frac{|f''|^2}{|f'|^2} - 3\frac{(f',f'')^2}{|f'|^4}\right]t^2 + \ldots$$

and finally

$$Sf(x,t) = |a,b,c,d| \left\{ 1 + \frac{1}{6}(a-b)(c-d)S_1 f(x)t^2 + o(t^2) \right\}. \tag{12}$$

This is the sought for asymptotic relation.

4. For the complex case there is a fairly well known geometric interpretation of $\operatorname{Im} Sf$ which reputedly goes back to G. Pick. Suppose that $z = f(t)$ represents a curve in the complex plane. The direction of its tangent is given by $\theta = \arg f'(t) = \operatorname{Im} \log f'(t)$. The curvature measures the rate of change of θ relative to arc length and is thus $K = |f'|^{-1}\operatorname{Im}(f''/f')$. One more differentiation shows that

$$dK/ds = |f'|^{-2}[\operatorname{Im}(f''/f')' - \operatorname{Re}(f''/f')\operatorname{Im}(f''/f')] = |f'|^{-2}\operatorname{Im} Sf(t). \tag{13}$$

We now imitate this computation for a curve $x = f(t)$ in R^n. The direction of the tangent is given by the unit vector $f'/|f'|$. Its derivative is

$$\frac{d}{dt}(f'/|f'|) = \frac{f''}{|f'|} - \frac{(f',f'')}{|f'|^3}f'$$

with the square norm

$$\left| \frac{d}{dt}(f'/|f'|) \right|^2 = \frac{|f''|^2}{|f'|^2} - \frac{(f',f'')^2}{|f'|^4} = \frac{|f' \wedge f''|^2}{|f'|^4}.$$

In other words, the curvature is $K = |f' \wedge f''|/|f'|^3$. Finally, the rate of change of the curvature is

$$dK/ds = \frac{(f' \wedge f'', f' \wedge f''')}{|f' \wedge f''||f'|^4} - 3\frac{|f' \wedge f''|(f',f'')}{|f'|^6}.$$

With the notation (8) this can be written in the form

$$\frac{dK}{ds} = |f'|^{-2}\frac{(S_2 f, f' \wedge f'')}{|f' \wedge f''|}. \tag{14}$$

The conclusion is that formula (13) remains in force provided that $\operatorname{Im} Sf$ is replaced by the projection of $S_2 f$ on the osculating plane.

Acknowledgements. I had never heard of complex distance until told by Troels Jørgensen, but I was familiar with this very natural notion without having a name for it. Subsequently I found it used in the famous unpublished manuscript of Fenchel and Nielsen. I had also access to an unpublished paper by W. Fenchel which I presume was a talk in Oberwolfach, and I read the original paper by F. Schilling. The old literature does not go beyond three dimensions.

References

[1] L. Ahlfors, Hyperbolic Motions, Nagoya Math. J. 28 (1967) 136–166.

[2] — , Invariant operators and integral representations in hyperbolic spaces. Mathematica Scandinavica 36 (1975) 27–43.

[3] — , A singular integral operator connected with quasiconformal mappings in space. L'Enseignement mathématique, t. XXIV, fasc. 3–4 (1978) 225–236.

[4] — , Möbius transformations in several dimensions. Ordway professorship lectures in mathematics, Univ. of Minnesota (1981).

[5] W. Fenchel, On trigonometry in hyperbolic 3-space. Unpublished.

[6] W. Fenchel and J. Nielsen, Discontinuous groups of non-euclidean motions. Unpublished.

[7] Fr. Schilling, Über die geometrische Bedeutung der Formeln in der sphärischen Trigonometrie im Falle complexer Argumente. Gött. Nachr. (1891). Reprinted in Crelle's Journal 1893.

Department of Mathematics
Harvard University
Cambridge
MA 02138
U.S.A.

J. M. Anderson, W. H. J. Fuchs

Remarks on "almost best" Approximation in the Complex Plane

§1

Let $f(x)$ be a continuous function on the compact interval J of the real axis, which is not the restriction of a function holomorphic in a neighborhood of J. Let π_n be the set of all polynomials over $\mathbb{C}$ of degree $\leq n$. Let $p_n(x)$ be the polynomial of best approximation to $f(x)$ on J, i.e.,

$$E_n(f, J) = \inf_{q \in \pi_n} \| f - q \|_J = \| f - p_n \|_J,$$

where the norm is the sup-norm.

R. Grothmann and E.B. Saff recently raised the following questions: [4]

(i) Can it happen that there is a function $F(z)$ defined in a domain D intersecting J and an $A > 0$ such that

$$\| F - p_n \|_D = \sup_{z \in D} |F(z) - p_n(z)| < A E_n(f, J) \quad (n \in \mathbb{N})?$$

(ii) Are there, at least, polynomials of "almost best" approximation $q_n(z)$ such that

$$\| f - q_n \|_J < A_1 E_n(f, J)$$
$$\| F - q_n \|_D < A_2 E_n(f, J) \, ?$$

In this paper we consider the example

$$J = [-a, a] \quad f(x) = |x|. \tag{1}$$

It is well known that in this case there are positive constants b and c such that

$$ab < n E_n(|x|, J) < ac. \tag{2}$$

We shall show

Proposition A. *In the case* (1), *no sub-sequence of* $\{p_n\}$ *converges uniformly in a domain D intersecting J.*

We also have

Theorem 1. *Let L be a compact subset of C with connected complement. Suppose that $f(z)$ is continuous on L and holomorphic in $L°$ but has a singularity on L. Then the sequence $\{p_n(z)\}$ does not converge in any domain intersecting L but not contained in $L°$.*

We are not able to exclude the possibility that a subsequence of $\{p_n\}$ might converge.

Proposition B. *Suppose that*

$$F(z) = z(Rz > 0), \quad F(z) = -z(Rz < 0), \quad F(0) = 0.$$

Let E be the union of an interval $[-a, a]$ of the real axis and of a compact set K not containing the origin but symmetric with respect to the origin and contained in $\{z : |\arg z| < \frac{\pi}{4}\} \cup \{z : |\arg z - \pi| < \frac{\pi}{4}\}$. Then one can find polynomials $q_n(z) \in \pi_n$ such that

$$\| F - q_n \|_E < A(E)/n \quad (n \in I\!N).$$

The proof of Proposition A and Theorem 1 is given in §2; the proof of Proposition B in §3. In §4 there are some further remarks and questions about the problems treated here.

§2

Proof of Prop. A. Let $p_n(x)$ be the polynomial of best approximation to $|x|$ on $[-1, 1]$. If $J = [-a, a]$, $q_n \in \pi_n$, then

$$\sup_{x \in J} |\,|x| - q_n(x)| = \sup_{|x| \le 1} a|\,|x| - (1/a)q_n(ax)|,$$

which shows that the polynomial of best approximation on J is given by $(1/a)q_n(ax) = p_n(x)$;

$$q_n(x) = ap_n(x/a).$$

It is therefore enough to consider the case $J = [-1, 1]$.

If in a domain D intersecting $J = [-1, 1]$

$$|F(z) - p_n(z)| < A/n,$$

then the sequence $\{p_n(z)\}$ converges uniformly in D and $F(z) = \lim p_n(z)$ is holomorphic in D. On $D \cap J$, $F(x) = x$ or $F(x) = -x$. It follows that either $F(z) = z$ or $F(z) = -z$ and that D intersects J either in a set on the positive real axis or in a set on the negative real axis. Consider the case that $F(z) = z$, the proof in the case $F(z) = -z$ is similar.

Let $d_n(z) = z - p_n(z)$. For given n, the points $u \in [-1, 1]$ at which $d_n(u) = E_n(1x1)$ and the points $v \in [-1, 1]$ at which $d_n(v) = -E_n(1x1)$ are interlaced and, by a result of S.N. Bernstein, ([2], p. 14) the distance of a "u" from the nearest "v" is less than A/n. If a subsequence $\{d_n(x) : n \in \Lambda\}$ converged uniformly to 0 in a domain intersecting $[0,1]$, then the sequence $\{d_n(x) : n \in \Lambda\}$ would converge uniformly to 0 in a subinterval K of $[0,1]$. But, for all large n, K would contain a "u" and a "v" less than A/n apart so that, by (2),

$$\frac{d_n(u) - d_n(v)}{u - v} > \frac{2E_n(1x1)}{A/n} > \frac{2b}{A}.$$

Rolle's Theorem now yields a contradiction.

Proof of Theorem 1. This proof is due to H.P. Blatt and E.B. Saff who prove in [3] that, for a subsequence $\{p_n : n \in \Lambda \subset N\}$, we have, outside a "small" exceptional set of z with empty interior

$$\limsup(1/n)|\log p_n(z)| = G(z) \qquad (z \notin L; \ n \to \infty, \ n \in \Lambda).$$

Here $G(z)$ is the Green's function with pole at ∞ of the complement of L in C. This obviously implies that, as $n \to \infty$, $p_n(z)$ cannot converge to a finite limit in a region containing points outside L, thus proving Theorem 1.

§3

Proof of Prop. B. Without loss of generality we may suppose that $a = 1$ and $K \subset L = \{z : |z^2 - 1| < 1\}$.

Suppose that K is not contained in L or that $a \neq 1$. Then one can find $\alpha > 0$ such that

$$\alpha K \subset L \quad \text{and} \quad \alpha a \leq 1.$$

If $E_1 = \alpha K \cup [-1, 1]$ and if

$$\| F - q_n \|_{E_1} < A_1/n,$$

then

$$\| F(\alpha z) - q_n(\alpha z) \|_E \leq A_1/n$$

and, since $F(\alpha z) = \alpha F(z)$,

$$\| F(z) - (1/\alpha)q_n(\alpha z) \| \leq A_1/(\alpha n) = A(E)/n.$$

Let

$$\zeta = z^2 - 1.$$

Then for real z in $|z| \leq 1$

$$|z| = \sqrt{1 + \zeta}.$$

Proposition B will be proved if we can establish the existence of $P_j \in \pi_j$ for all $j > j_0$ such that

$$|\sqrt{1+\zeta} - P_j(\zeta)| < A(\delta)/j \quad (-1 \leq \zeta \leq 1) \tag{3}$$

$$|\sqrt{1+\zeta} - P_j(\zeta)| < A(\delta)/j \quad (|\zeta| \leq 1 - \delta). \tag{4}$$

Let

$$D(\zeta) = \sqrt{1+\zeta} - \sum_{k=0}^{n-1} \begin{bmatrix} 1/2 \\ k \end{bmatrix} \zeta^k$$

$$= \sum_{k=n}^{\infty} \begin{bmatrix} 1/2 \\ k \end{bmatrix} \zeta^k \quad (|\zeta| < 1).$$

Using

$$\frac{\Gamma(p)\Gamma(q)}{\Gamma(p+q)} = \int_0^1 t^{p-1}(1-t)^{q-1}dt \quad (p, q > 0)$$

and

$$\Gamma(z)\Gamma(1-z) = \frac{\pi}{\sin \pi z}$$

we have

$$\begin{bmatrix} 1/2 \\ k \end{bmatrix} = \frac{1}{2} \cdot \left(\frac{1}{2} - 1\right)\left(\frac{1}{2} - 2\right) \cdot \left(\frac{1}{2} - k + 1\right) \Big/ k!$$

$$= (-1)^k \frac{\Gamma(k - \frac{1}{2})}{\Gamma(-\frac{1}{2})\Gamma(k+1)}$$

$$= (-1)^k \frac{1}{\Gamma(-\frac{1}{2})\Gamma(\frac{3}{2})} \frac{\Gamma(\frac{3}{2})\Gamma(k - \frac{1}{2})}{\Gamma(k+1)}$$

$$= (-1)^k \frac{\sin(-\pi/2)}{\pi} \int_0^1 (1 - t)^{1/2} t^{k-3/2} dt$$

$$= (-1)^{k-1}(1/\pi) \int_0^1 (1 - t)^{1/2} t^{k-3/2} dt.$$

Therefore

$$\left. \begin{aligned} D(\zeta) &= -(1/\pi) \int_0^1 (1 - t)^{1/2} \sum_{k=n}^{\infty} t^{k-3/2}(-\zeta)^k dt \quad (|\zeta| < 1) \\[2mm] &= -(1/\pi)(-\zeta)^n \int_0^1 (1 - t)^{1/2} t^{n-3/2}(1 + \zeta)^{-1} dt. \end{aligned} \right\} \tag{5}$$

From this expression it is easy to see that the choice

$$P_{n-1}(\zeta) = \sum_{k=0}^{n-1} \begin{bmatrix} 1/2 \\ k \end{bmatrix} \zeta^k$$

satisfies (4). But this choice does not satisfy (3) since

$$|D(-1)| = \frac{1}{\pi} \int_0^1 (1 - t)^{-1/2} t^{n-3/2} dt = \frac{\Gamma(\frac{1}{2})\Gamma(n - \frac{1}{2})}{\pi \Gamma(n)} \sim \text{const.} \, n^{-1/2}.$$

We try to improve our choice of P by setting

$$P_{n+m}(\zeta) = P(\zeta) = \sum_0^{n-1} \begin{bmatrix} 1/2 \\ k \end{bmatrix} \zeta^k - (-\zeta)^n Q_m(\zeta)$$

where

$$Q_m(\zeta) = \frac{1}{\pi} \int_{2/3}^{1} (1-t)^{1/2} t^{n-3/2} q_t(\zeta) dt, \quad q_t \in \pi_m.$$

Thus, by (5),

$$\left.\begin{aligned}
\sqrt{1+\zeta} - P_{n+m}(\zeta) &= (-1)^{n-1}(1/\pi)\zeta^n \int_0^{2/3} (1-t)^{1/2} t^{n-3/2} (1+t\zeta)^{-1} dt \\
&\quad + (-1)^{n-1}(1/\pi)\zeta^n \int_{2/3}^{1} (1-t)^{1/2} t^{n-3/2} \left((1+t\zeta)^{-1} - q_t(\zeta)\right) dt \\
&= I_1 + I_2.
\end{aligned}\right\} \quad (6)$$

For $|\zeta| \le 1$ we have $|(1+t\zeta)^{-1}| \le 3$ in I_1 and so

$$\left.\begin{aligned}
|I_1| &< |\zeta|^n \int_0^1 (1-t)^{1/2} t^{n-3/2} dt = |\zeta|^n \frac{\Gamma(\frac{3}{2})\Gamma(n-\frac{1}{2})}{\Gamma(n+1)} \\
&< \text{const.} \, |\zeta|^n n^{-3/2}.
\end{aligned}\right\} \quad (7)$$

We now choose for q_t the polynomial of best approximation of degree $\le m$ to $(1+t\zeta)^{-1}$ in $-1 \le \zeta \le 1$. Then

$$E_t = E_m(1/(1+t\zeta), [-1,1]) = \sup_{-1 \le \zeta \le 1} t^{-1} \left| \frac{1}{(1/t) + \zeta} - tq_t(\zeta) \right|.$$

Chebyshev found explicit formulas for E_t and tq_t (see [1], p.59 and 60) which yield

$$E_t = \frac{t^{m+1}}{(1-t^2)\left(1 + \sqrt{1-t^2}\right)^m} \quad (8)$$

and

$$tq_t(\zeta) = \frac{1}{(1/t) + \zeta} + \frac{tE_t}{2} \left\{ v^m \frac{s-v}{1-sv} + v^{-m} \frac{1-sv}{s-v} \right\}, \quad (9)$$

where

$$-\zeta = \frac{1}{2}\left(v + \frac{1}{v}\right), \quad s = t/1 + \sqrt{1-t^2}.$$

By (6) and (8) for $-1 \leq \zeta \leq 1$

$$\left.\begin{aligned}
|I_2| &\leq |\zeta|^n (1/\pi) \int_{2/3}^{1} (1-t)^{1/2} t^{n-3/2} E_t dt \\[2ex]
&\leq |\zeta|^n (1/n) \int_{2/3}^{1} (1-t^2)^{-1/2} (1+t)^{-1/2} t \left(1 + \sqrt{1-t^2}\right)^{-m} dt \\[2ex]
&\leq |\zeta|^n (1/n) \int_{2/3}^{1} (1-t^2)^{-1/2} t \left(1 + \sqrt{1-t^2}\right)^{-m} dt \\[2ex]
&\leq |\zeta|^n (1/\pi) \int_{0}^{5/9} (1+u)^{-m} du \leq \frac{A}{m-1} |\zeta|^n \leq \frac{A}{m-1}.
\end{aligned}\right\} \qquad (10)$$

(6), (7) and (10) prove (3), provided that for some positive α

$$m > \alpha n.$$

It remains to prove (4), or, equivalently, to show that in $|\zeta| \leq 1$, $|I_2| < A_1/(n+m)$. We estimate $q_t(\zeta)$ for $|\zeta| = 2$, $2/3 \leq t < 1$. By (9)

$$|q_t(\zeta)| \leq \frac{F_t}{2} \left| v^m \frac{s-v}{1-sv} + v^{-m} \frac{1-sv}{s-v} \right| + \left| \frac{1}{1+t\zeta} \right|. \qquad (11)$$

The second term on the right is bounded by 3 for the values of ζ and t under consideration. For $|\zeta| = 2$ the equation $\zeta = -\frac{1}{2}(v + v^{-1})$ has a single root v in $|v| > 1$ for every ζ on $|\zeta| = 2$. As ζ describes the circle $|\zeta| = 2$, v describes a curve whose minimum and maximum distance from $v = 0$ are given by

$$\frac{1}{2}\left(|v|_{\min} + (1/|v|_{\min})\right) = 2$$

$$\frac{1}{2}\left(|v|_{\max} - (1/|v|_{\max})\right) = 2,$$

i.e., $v_{\min} = 2 + \sqrt{3} < 3.8$ and $v_{\max} = 2 + \sqrt{5}$.

The definition of s (see (9)) shows that for $\frac{2}{3} \leq t \leq 1$

$$.38 < \frac{2}{3} \cdot \frac{1}{1 + \sqrt{5/9}} < s < 1.$$

Therefore, for $|\zeta| = 2$,

$$\left| \frac{s - v}{1 - sv} \right| < \frac{1 + |v|_{\max}}{.38|v|_{\min} + 1} < \infty$$

$$\left| \frac{1 - sv}{s - v} \right| < \frac{1 + |v_{\max}|}{|v|_{\min} - 1} < \infty.$$

By (11) and the maximum modulus principle it follows now that

$$|q_t(\zeta)| \leq AE_t \cdot (2 + \sqrt{5})^m + 3 \quad \left(|\zeta| \leq 2, \frac{2}{3} \leq t < 1 \right).$$

Therefore in $|\zeta| \leq 1 - \delta$

$$\pi|I_2| \leq |\zeta|^n \int_{2/3}^{1} (1 - t)^{1/2} t^{n-3/2} |(1 + t\zeta)^{-1}| dt$$

$$+ A|\zeta|^n \int_{2/3}^{1} (1 - t)^{1/2} t^{n-3/2} E_t (2 + \sqrt{5})^m dt$$

$$+ 3|\zeta|^n \int_{2/3}^{1} (1 - t)^{1/2} t^{n-3/2} dt.$$

Using $|(1 + t\zeta)^{-1}| \leq 1/\delta$ and (10) we have

$$\pi|I_2| \leq (1 - \delta)^w \int_{0}^{1} (1 - t)^{1/2} t^{n-1/2} \{(1/\delta) + 3\}$$

$$+ A(\delta)(1 - \delta)^n (2 + \sqrt{5})^m / (m - 1)$$

$$\leq A(\delta)(1 - \delta)^n (2 + \sqrt{5})^m \left\{ \frac{1}{(m - 1)} + n^{-3/2} \right\}.$$

If m is chosen in the range

$$\frac{1}{2} \left(\log \left(\frac{1}{1 - \delta} \right) \Big/ \log(2 + \sqrt{5}) \right) n < m < \left(\log \left(\frac{1}{1 - \delta} \right) \Big/ \log(2 + \sqrt{5}) \right) n,$$

$$(12)$$

then (3) and (4) hold with $j = n + m$. Since m is subject only to the conditions (12), we can choose n and m so that $n + m$ runs through all sufficiently large integers.

§4

We should like to point out several problems related to the preceding remarks.

1) Can Proposition B be improved by replacing E by $[-a, a] \cup K \cup (-K)$ where K is a compact set in $x > 0$? What about the possibility that K also contains the point $z = 0$? Grothmann and Saff [4] conjecture that K could not contain an angle $|\arg z| \leq \alpha$, $|z| \leq \delta$, but that it is possible that K contains the region $y \leq Ax^2$, $0 \leq x \leq \delta$ $(A > 0)$.

2) What is the generalization of Proposition B, if $|x|$ is replaced by an $f(x)$ satisfying the conditions stated in §1?

Bibliography

[1] N.I. Achieser, Vorlesungen über Approximationstheorie. Akademie Verlag, Berlin, 1953.

[2] S.N. Bernstein, "Sur la meilleure approximation de $|x|$ par les polynomes de degrés donnés". Acta M. 37 (1914), 1–57.

[3] H.P. Blatt and E.B. Saff, "Behavior of Zeros of Polynomials of Near Best Approximations". Submitted to Math. Ztschr.

[4] R. Grothmann and E.B. Saff, "On the behavior of zeros and poles of best uniform polynomial and rational approximants", to appear.

Mathematics Department
University College London
Gower Street
London WC1E 6BT, U.K.

Mathematics Department
Cornell University
Ithaca
N.Y. 14853, U.S.A.

P. L. Duren, M. M. Schiffer

Conformal Mappings onto Nonoverlapping Regions

§1 Introduction

Let $f(\zeta) = a + d\zeta \ldots$ be analytic and univalent in the unit disk $|\zeta| < 1$, mapping it conformally onto some domain D. We shall call $a = f(0)$ the *center* and $|d| = |f'(0)|$ the *inner radius* of D with respect to a. Roughly speaking, our problem is to find n functions

$$f_j(\zeta) = a_j + d_j\zeta + \ldots, \quad j = 1, 2, \ldots, n, \tag{1}$$

which map the disk conformally onto nonoverlapping regions D_j whose union has prescribed transfinite diameter R, with the centers a_j as far apart as possible and the inner radii $|d_j|$ as large as possible. Here only n and R are specified in advance.

A suitable statement of this problem, in generalized form, is as follows. Fix nonzero real parameters $x_1, x_2, \ldots, x_n$ and ask for the maximum value of the functional

$$\phi = \sum_{\substack{j,k=1 \\ j \neq k}}^{n} x_j x_k \log |a_j - a_k| + \sum_{j=1}^{n} x_j^2 \log |d_j| \tag{2}$$

among all systems of functions $f_1, f_2, \ldots, f_n$ satisfying the given conditions. Using a variational method, we shall establish the sharp inequality

$$\phi \leq s^2 \log R, \quad \text{where} \quad s = \sum_{j=1}^{n} x_j. \tag{3}$$

Special cases of the inequality (3) were previously found by Alenicyn [1] and Kühnau [4], whose methods were quite different from ours. The variational method is particularly well adapted to the functional (2) because it leads to a perfect square in the associated quadratic differential.

It is not immediately clear that an extremal configuration exists. In order to handle this difficulty, or rather to avoid it, we shall begin by considering a different form of the extremal problem. Specifically, we shall require that the regions D_j lie in the complement of a fixed domain Ω in the extended complex plane $\hat{\mathcal{C}}$, or in the complement of some domain conformally equivalent to Ω under a normalized mapping of the form

$$w = z + \sum_{k=0}^{\infty} b_k z^{-k}. \tag{4}$$

We first prove (3) under the assumption that Ω is simply connected. In the case where all $x_j = 1$, we then exhibit a specific extremal configuration. Next allowing Ω to be multiply connected, we apply the same method to prove a stronger form of the inequality (3). Here the analysis of the quadratic differential is more difficult and it is necessary to consider harmonic measures of the boundary components and the period matrix of their harmonic conjugates.

§2 Simply connected complements

Let $\Omega \subset \hat{\mathcal{C}}$ be a simply connected domain containing the point at infinity, with complement $\tilde{\Omega}$ of given transfinite diameter $R > 0$. In view of the Riemann mapping theorem, it is equivalent to suppose that Ω is the conformal image of a fixed domain of this type under an arbitrary mapping of the form (4). Assuming as we may that $\tilde{\Omega}$ has an interior, we consider a set of n functions f_j of the form (1), analytic and univalent in $|\zeta| < 1$, with disjoint ranges D_j all contained in $\tilde{\Omega}$. Holding n fixed, we wish to choose Ω and the f_j to maximize the functional ϕ given in (2).

By the Koebe one-quarter theorem and a compactness argument, it is easily seen that an extremal configuration exists if no subset of the parameters x_j has sum zero. The existence of an extremal configuration under a fixed noncritical choice of parameters will allow us to apply a variational method to establish the sharp inequality (3). An obvious continuity argument then extends (3) to an arbitrary system of parameters x_j.

We again use the notation Ω, f_j, and D_j to indicate an extremal configuration. Fix a point z_0 in

$$\Gamma = \tilde{\Omega} \cap \bigcap_{j=1}^{n} \tilde{D}_j$$

and construct the boundary variation [6, 2]

$$z^* = V(z) = z + \frac{a\rho^2}{z - z_0} + O(\rho^3).$$
(5)

This function V is analytic and univalent outside a small part of Γ near z_0. In particular, the functions

$$f_j^*(\zeta) = V(f_j(\zeta)) = a_j^* + d_j^*\zeta + \dots$$

are again univalent and map the disk onto nonoverlapping regions $D_j^* = V(D_j)$ contained in $\tilde{\Omega}^*$, where $\Omega^* = V(\Omega)$.

Furthermore, the variation preserves the transfinite diameter. This is the best seen by considering Green's function

$$g(z, \infty) = \log|z| + \gamma + u(z)$$
(6)

of the region Ω. Here u is harmonic in Ω with $u(\infty) = 0$. It is well known (see [3], Ch. VII, §3) that the transfinite diameter of $\tilde{\Omega}$ is $R = e^{-\gamma}$, where Robin's constant γ is defined by (6). But Green's function of Ω^* is $g^*(z^*, \infty) = g(z, \infty)$, and a simple calculation shows that $\gamma^* = \gamma$.

Turning now to the variation of the functional ϕ, we find

$$a_j^* = f_j^*(0) = a_j - a\rho^2(z_o - a_j)^{-1} + O(\rho^3);$$

$$d_j^* = f_j^{*\prime}(0) = d_j\left[1 - a\rho^2(z_o - a_j)^{-2} + O(\rho^3)\right].$$

Thus a short calculation gives

$$\phi^* = \phi - \mathrm{Re}\left\{a\rho^2\left(\sum_{j=1}^n \frac{x_j}{z_o - a_j}\right)^2 + O(\rho^3)\right\}.$$

Using the inequality $\phi^* \leq \phi$ and invoking the basic lemma of the theory of boundary variation, we conclude that the points $z_o \in \Gamma$ lie on trajectories of the quadratic differential

$$-\left(\sum_{j=1}^n \frac{x_j}{z - a_j}\right)^2 dz^2 > 0.$$
(7)

Because of the perfect square, the quadratic differential (7) is easily analyzed. Introducing a parametrization $z = z(t)$, one takes the square root and integrates to obtain the equation

$$\sum_{j=1}^{n} x_j \log|z - a_j| = C \tag{8}$$

for an arc of Γ, where C is a real constant. Now introduce Green's function

$$g(z, \infty) = \log|z| - \log R + o(1), \quad z \to \infty,$$

of Ω and observe that

$$\sum_{j=1}^{n} x_j \log|z - a_j| - C = s \log|z| - C + O(1/z) = s\, g(z, \infty),$$

where $s = \sum_{j=1}^{n} x_j$. Thus $C = s \log R$, and exponentiation of (8) gives

$$\prod_{j=1}^{n} |z - a_j|^{x_j} = R^s. \tag{9}$$

Thus the regions D_j fill $\tilde{\Omega}$ and their boundary arcs lie on the lemniscate (9).

The equation (8) shows that for each k $(1 \leq k \leq n)$ the equation

$$\sum_{j=1}^{n} x_j \log|f_k(\zeta) - a_j| - x_k \log|\zeta| - C = 0$$

holds on $|\zeta| = 1$; hence by the maximum principle this expression vanishes identically in $|\zeta| < 1$. (Note that the subtraction of $x_k \log|\zeta|$ removes the singularity at the origin.) Choosing $\zeta = 0$, we obtain

$$\sum_{\substack{j=1 \\ j \neq k}}^{n} x_j \log|a_j - a_k| + x_k \log|d_k| = s \log R.$$

Multiplication by x_k and summation over k therefore gives $\phi = s^2 \log R$ for the extremal configuration. It follows that the inequality (3) holds for all admissible configurations.

In summary, we have proved the following theorem.

Theorem 1. *Let $\Omega \subset \hat{\mathbb{C}}$ be a simply connected domain containing infinity, with complement $\tilde{\Omega}$ of transfinite diameter R. Let $f_1, f_2, \ldots, f_n$ be analytic functions of the form (1) which map the unit disk conformally onto nonoverlapping regions D_j contained in $\tilde{\Omega}$. Choose real parameters $x_1, x_2, \ldots, x_n$ with sum s and define the functional ϕ as in (2). Then the sharp inequality $\phi \leq s^2 \log R$ holds. When equality occurs, the regions D_j fill $\tilde{\Omega}$ (leaving no open set uncovered), and their boundary points satisfy (9).*

It may be remarked that although the bound is always sharp, it need not be attained if a subset of the parameters x_j has sum zero. For example, let $n = 2$ and choose $x_1 = 1$, $x_2 = -1$. Then if an extremal configuration exists, the above argument shows that the boundaries of D_1 and D_2 both satisfy (9), which now becomes

$$\left| \frac{z - a_1}{z - a_2} \right| = 1.$$

But this is a line separating a_1 and a_2, so D_1 and D_2 are half-planes, and they cannot be contained in the bounded region $\tilde{\Omega}$.

§3 An extremal configuration

Guided by the equation (9), we now produce an example of an extremal configuration for a particular case. Let us take all $x_j = 1$, and choose $a_j = R\omega^j$, where $\omega = e^{2\pi i/n}$ is a primitive n^{th} root of unity. Then (9) takes the form

$$\prod_{j=1}^{n} |z - a_j| = R^n,$$

or equivalently $|z^n - R^n| = R^n$. Its locus is a system of lemniscates as shown in Figure 1. The polynomial $p(z) = (z/R)^n - 1$ maps each petal D_j conformally onto the unit disk $|\zeta| < 1$, with $p(a_j) = 0$. Thus the local inverse $z = f_j(\zeta)$, with $f_j(0) = a_j$, maps the disk conformally onto D_j. Because

$$p'(a_j) = n R^{-n} a_j^{n-1} = n R^{-1} \omega^{-j},$$

we have

$$d_j = f_j'(0) = R\omega^j/n, \quad j = 1, 2, \ldots, n.$$

The functional ϕ therefore takes the value

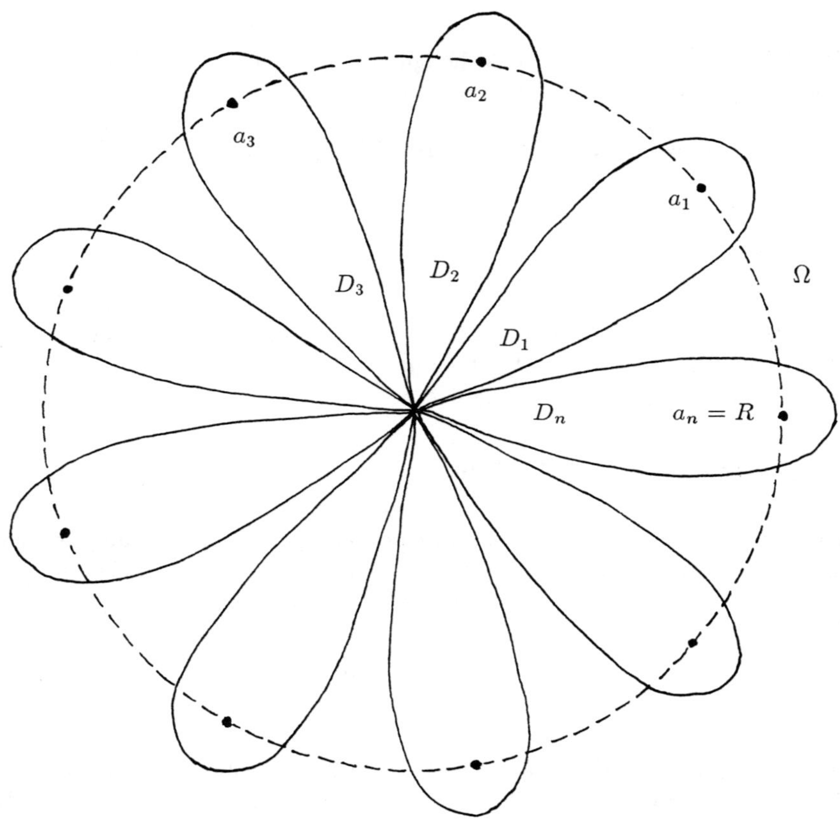

Figure 1. An extremal configuration

$$\phi = n(n-1)\log R + \sum_{\substack{j,k=1 \\ j \neq k}}^{n} \log|\omega^j - \omega^k| + \sum_{j=1}^{n} \log|d_j|$$

$$= n^2 \log R - n\log n + \sum_{\substack{j,k=1 \\ j \neq k}}^{n} \log|\omega^j - \omega^k|. \tag{10}$$

But for each fixed k the polynomial

$$q(w) = \prod_{j=1}^{n}(w - \omega^j) = w^n - 1$$

has the derivative

$$q'(\omega^k) = \prod_{\substack{j=1 \\ j \neq k}}^{n}(\omega^k - \omega^j) = n\omega^{k(n-1)} = n\omega^{-k}. \tag{11}$$

Introducing the identity (11) into (10), we find that $\phi = n^2 \log R$, the maximum value. Thus the given configuration is extremal.

It remains to verify that the configuration is admissible; *i.e.*, that $\tilde{\Omega} = \bigcup_{j=1}^{n} \overline{D_j}$ has transfinite diameter R. But because the lemniscate $\partial\Omega$ has the equation $|z^n - R^n| = R^n$, we may simply observe that Green's function of Ω is

$$g(z, \infty) = \log|z| - \log R + \frac{1}{n}\log|1 - (R/z)^n|.$$

Thus the Robin constant for Ω is $-\log R$, which shows that R is the transfinite diameter of $\tilde{\Omega}$.

It is an open question whether this is the only extremal configuration, up to rigid motion, in the case where all $x_j = 1$.

§4 Multiply connected complements

We now generalize the extremal problem solved above by allowing Ω to be multiply connected. Suppose that its complement $\tilde{\Omega}$ is the union of m disjoint compact connected sets $\Delta_1, \Delta_2, \ldots, \Delta_m$, all containing open sets. We regard Ω as the conformal image of some fixed domain of this type under a mapping of the form (4). Let Ω have transfinite diameter R, invariant under conformal mappings of the form (4). Choose an integer $n \geq m$, and fix real numbers $x_1, x_2, \ldots, x_n$. Again consider n functions of the form (1), analytic and univalent in the unit disk with disjoint ranges D_j such that $D_j \subset \Delta_k$ for $n_{k-1} < j \leq n_k$, where $n_0 = 0$ and $n_m = n$. The problem is again to maximize the functional ϕ defined in (2) by choosing the domain Ω and the functions f_j subject to the above requirements. Here the numbers n_k are held fixed.

The existence of an extremal configuration is proved as in the simply connected case. As before, we must initially avoid the case where some of the x_j in the same block ($n_{k-1} < j \leq n_k$) have sum zero. Using the method of boundary variation in exactly the same way, we find that in the extremal situation the domains D_j fill Δ_k for each k ($n_{k-1} < j \leq n_k$), and their boundaries ∂D_j are trajectories of the quadratic differential (7). Integration now gives

$$\sum_{j=1}^{n} x_j \log |z - a_j| = C_k, \quad k = 1, 2, \ldots, m, \tag{12}$$

for $z \in \partial D_j \subset \overline{\Delta_k}$. In particular, the boundary Γ_k of Δ_k is a Jordan curve satisfying (12).

To interpret this result properly, we have to recall some standard concepts of potential theory (see Nehari [5], Ch. 1). Let $\omega_k(z)$ be the harmonic measure of Γ_k with respect to Ω. Let

$$P_{kj} = \frac{1}{2\pi} \int_{\Gamma_j} \frac{\partial \omega_k}{\partial n} |dz|$$

be the period of the harmonic conjugate of ω_k around Γ_j. Here $\partial/\partial n$ denotes the inner normal derivative. Then $P_{jk} = P_{kj}$ and the $(m-1) \times (m-1)$ matrix $((P_{jk}))$ is nonsingular, where $j, k = 1, 2, \ldots, m - 1$. In fact, it generates a negative-definite quadratic form. Let $((p_{jk}))$ denote the inverse matrix.

Now let $g(z, \infty)$ be Green's function of Ω with pole at ∞, and consider

$$F(z) = \sum_{j=1}^{n} x_j \log |z - a_j|, \quad z \in \Omega. \tag{13}$$

Let $s = \sum_{j=1}^{n} x_j$ and observe that

$$F(z) = s \log |z| + O(1/|z|), \quad z \to \infty.$$

Hence in view of (12), the function

$$F(z) - s \, g(z, \infty) - \sum_{k=1}^{m} C_k \omega_k(z)$$

is harmonic in Ω (even at ∞) and vanishes identically on each boundary curve Γ_k. It follows that

$$F(z) = s\,g(z,\infty) + \sum_{k=1}^{m} C_k \omega_k(z). \tag{14}$$

At $z = \infty$ this gives

$$\sum_{k=1}^{m} C_k \omega_k(\infty) = s \log R. \tag{15}$$

Next let

$$\sigma_k = \sum_{n_{k-1} < j \leq n_k} x_j \tag{16}$$

and use (13) and (14) to calculate

$$\sigma_k = \frac{1}{2\pi} \int_{\Gamma_k} \frac{\partial F}{\partial n} |dz| = s\,\omega_k(\infty) + \sum_{j=1}^{m} C_j P_{jk}.$$

Since $\sum_{j=1}^{m} P_{jk} = 0$, this may be written

$$\sum_{j=1}^{m-1} (C_m - C_j) P_{jk} = s\,\omega_k(\infty) - \sigma_k.$$

We now use the inverse matrix $((p_{jk}))$ to arrive at

$$C_m - C_j = \sum_{k=1}^{m-1} p_{kj} \left[s\,\omega_k(\infty) - \sigma_k \right], \quad j = 1, 2, \ldots, m-1. \tag{17}$$

Similarly, the identity $\omega_1(z) + \ldots + \omega_m(z) \equiv 1$ allows us to write (15) in the form

$$C_m + \sum_{k=1}^{m-1} (C_k - C_m) \omega_k(\infty) = s \log R. \tag{18}$$

Combining (17) and (18), we have

$$C_m = s \log R + \sum_{j=1}^{m-1} \sum_{k=1}^{m-1} p_{jk} \left[s\,\omega_j(\infty) - \sigma_j \right] \omega_k(\infty). \tag{19}$$

The formulas (17) and (19) evaluate the constants C_k $(1 \leq k \leq m)$ in terms of the domain Ω.

We now turn to the evaluation of ϕ for the extremal configuration. From (12) we conclude that

$$\sum_{j=1}^{n} x_j \log |f_i(\zeta) - a_j| - x_i \log |\zeta| = C_k$$

on $|\zeta| = 1$ for $n_{k-1} < i \leq n_k$, hence also in $|\zeta| < 1$. Putting $\zeta = 0$, we find

$$\sum_{\substack{j=1 \\ j \neq i}}^{n} x_j \log |a_i - a_j| + x_i \log |d_i| = C_k.$$

Multiplying by x_i and summing over i, we arrive at

$$\phi = \sum_{k=1}^{m} \sigma_k C_k.$$

Finally, we insert the expressions (17) and (19) for the C_k to obtain after minor rearrangement

$$\phi = \sum_{k=1}^{m-1} \sigma_k (C_k - C_m) + s\, C_m$$

$$= s^2 \log R + \sum_{j=1}^{m-1} \sum_{k=1}^{m-1} p_{jk} \left[s\, \omega_j(\infty) - \sigma_j \right] \left[s\, \omega_k(\infty) - \sigma_k \right].$$

We have calculated the value of ϕ in the maximum case and have therefore established a sharp inequality. In particular, we have proved the following theorem.

Theorem 2. *Let $\Omega \subset \hat{\mathcal{C}}$ be a finitely connected domain whose boundary consists of Jordan curves Γ_k with bounded interiors Δ_k $(k = 1, 2, \ldots, m)$ comprising a set of transfinite diameter R. Choose integers $n \geq m$ and n_k with $1 < n_1 < n_2 < \ldots < n_m = n$. Let x_j be positive numbers with sum s, and define σ_k as in (16). Let the functions f_j $(j = 1, 2, \ldots, n)$ have the form (1) and map the unit disk conformally onto nonoverlapping domains D_j with $D_j \subset \Delta_k$ for $n_{k-1} < j \leq n_k$. Then the functional ϕ defined by (2) satisfies the sharp inequality*

$$\phi \leq s^2 \log R + \sum_{j=1}^{m-1} \sum_{k=1}^{m-1} p_{jk} \left[s\, \omega_j(\infty) - \sigma_j \right] \left[s\, \omega_k(\infty) - \sigma_k \right],$$

where ω_k is the harmonic measure of Γ_k and $((p_{jk}))$ is the inverse of the period matrix of Ω.

Corollary 1. *Under the assumptions of the theorem, the functional ϕ satisfies the sharp inequality $\phi \le s^2 \log R$.*

Proof. The matrix $((p_{jk}))$ generates a negative-definite quadratic form and so the inequality (20) clearly implies $\phi \le s^2 \log R$. This weaker inequality is certainly sharp for a suitable choice of the x_j, given Ω and the corresponding (conformally invariant) quantities $\omega_k(\infty)$. Indeed, we need only choose the x_j such that $\sigma_k = s\,\omega_k(\infty)$ for each k. However, it is actually sharp for each choice of x_j with all $\sigma_k > 0$, because Ω can be constructed with prescribed (positive) invariants $\omega_k(\infty)$ whose sum is 1.

This last assertion can be proved as follows. Given numbers $q_j > 0$ with $\sum_{j=1}^{n} q_j = 1$, choose n different points a_j and consider the lemniscate

$$\prod_{j=1}^{n} |z - a_j|^{q_j} = c.$$

Let the positive constant c be small enough so that the lemniscate separates into n simple closed curves Γ_j about the points a_j. Let Ω be the region outside of the curves Γ_j. Then Green's function of Ω is

$$g(z, \infty) = \sum_{j=1}^{n} q_j \log |z - a_j| - \log c.$$

Now

$$\omega_k(\infty) = \frac{1}{2\pi} \int_{\Gamma_k} \frac{\partial}{\partial n} g(z, \infty) |dz|$$

$$= \sum_{j=1}^{n} q_j \frac{1}{2\pi} \int_{\Gamma_k} \frac{\partial}{\partial n} \log |z - a_j| \, |dz|.$$

But by Gauss' theorem,

$$\frac{1}{2\pi} \int_{\Gamma_k} \frac{\partial}{\partial n} \log |z - a_j| \, |dz| = \delta_{jk}.$$

Thus $\omega_k(\infty) = q_k$ for $k = 1, 2, \ldots, n$. $\square$

We remark that Shih [7] has obtained a more general result in the context of Brownian motion.

Corollary 2. *Let n functions f_j map the unit disk conformally onto nonoverlapping regions D_j whose union has transfinite diameter R. Then $\phi \leq s^2 \log R$.*

§5 Applications

First consider two functions f_j with disjoint ranges D_j, centers a_j, and inner radii $|d_j|$. Choose $x_1 = 1$ and $x_2 = -1$. Then $s = 0$ and so $\phi \leq 0$. Specifically,

$$-2\log|a_1 - a_2| + \log|d_1| + \log|d_2| \leq 0,$$

or

$$|d_1 d_2| \leq |a_1 - a_2|^2.$$

If D_1 and D_2 have disjoint closures, Theorem 2 may be applied to yield

$$\frac{1}{P_{11}} = p_{11} \geq \log\left|\frac{d_1 d_2}{(a_1 - a_2)^2}\right|.$$

But it is easily seen that $P_{11} = -(\log M)^{-1}$, where M is the modulus of the doubly connected domain Ω exterior to D_1 and D_2. (Here M is the ratio of the radii of a conformally equivalent annulus.) Thus we have the sharp inequality

$$M|d_1 d_2| \leq |a_1 - a_2|^2.$$

Next consider n functions f_j and choose all $x_j = 1$. Then the inequality $\phi \leq s^2 \log R$ becomes

$$2\sum_{j<k} \log|a_j - a_k| + \sum_{j=1}^{n} \log|d_j| \leq n \log R,$$

or

$$\prod_{j=1}^{n} |d_j| \prod_{j<k} |a_j - a_k|^2 \leq R^n.$$

If in particular the f_j are nonoverlapping translations of a given mapping $f(\zeta) = d\zeta + \dots$ and their centers a_j lie at the vertices of a regular n-gon with sidelength L, then we have

$$|d|L^{n-1} \leq R.$$

This may be viewed as an estimate for the transfinite diameter. We obtain the same estimate more generally under the assumption that the n translations f_j have centers a_j with distance no less than L from each other.

If $R \leq 1$, the inequality $\phi \leq s^2 \log R$ may be interpreted as implying that a certain quadratic form generated by centers and inner radii is negative semidefinite.

We mention in closing that the functional on the right-hand side of the inequality (20) is a very interesting combination which also has a perfect square in its variation. We hope to amplify this remark on another occasion.

Acknowledgement. The research of the first-named author was supported in part by the National Science Foundation under Grant DMS-8701751.

References

[1] Yu. E. Alenicyn, "On univalent functions in multiply connected domains", Mat. Sb. 39 (81) (1956), 315–336. (in Russian)

[2] P. L. Duren, *Univalent Functions* (Springer-Verlag, Heidelberg and New York, 1983).

[3] G. M. Goluzin, *Geometric Theory of Functions of a Complex Variable* (Moscow, 1952; German transl., Deutscher Verlag, Berlin, 1957; 2nd ed., Moscow, 1966; English transl., Amer. Math. Soc., 1969).

[4] R. Kühnau, "Über die schlichte konforme Abbildung auf nichtüberlappende Gebiete", Math. Nachr. 36 (1968), 61–71.

[5] Z. Nehari, *Conformal Mapping* (McGraw-Hill, New York, 1952).

[6] M. Schiffer, "A method of variation within the family of simple functions", Proc. London Math. Soc. 44 (1938), 432–449.

[7] C.-T. Shih, to appear.

Department of Mathematics
University of Michigan
Ann Arbor, Michigan 48109
U.S.A.

Department of Mathematics
Stanford University
Stanford, California 94305
U.S.A.

Matts Essén

On Wiener Conditions for minimally thin and rarefied Sets

§1 Introduction

Let $D = \{x \in I\!R^p : x_1 > 0\}$, where $x = (x_1, \ldots, x_p)$, $p \geq 2$ and ∂D is the euclidean boundary of D. If u is subharmonic in D and $y \in \partial D$, we define $u(y) = \limsup u(x)$, $x \to y$, $x \in D$. If u is non-positive on ∂D and $\sup_D u(x)/x_1 < \infty$, it is known that

$$u(x)/x_1 \to \alpha, \quad x \to \infty, \quad x \in D \setminus E,$$

$$(u(x) - \alpha x_1)/|x| \to 0, \quad x \to \infty, \quad x \in D \setminus F,$$

where the exceptional set E is minimally thin at infinity in D (cf. [5]) and the exceptional set F is rarefied at infinity in D (cf. [3]).

These exceptional sets can be characterized in terms of generalized Wiener conditions. Let $\gamma(E)$ be the Green energy of a set $E \subset D$ and let $\lambda'(E)$ be the Green mass of a set $E \subset D$ (definitions will be given below). Then the sets E and F are minimally thin or rarefied at infinity in D, respectively, if and only if

$$\sum_1^\infty \gamma(E^{(n)})2^{-np} < \infty, \tag{1.1}$$

$$\sum_1^\infty \lambda'(F^{(n)})2^{n(1-p)} < \infty, \tag{1.2}$$

where $E^{(n)}$ and $F^{(n)}$ are the intersections of E and F with the half-annulus $\{x \in D \cup \partial D : 2^n \leq |x| < 2^{n+1}\}$, $n = 1, 2, \ldots$.

Can these Wiener conditions be written in terms of the ordinary (euclidean) capacities of the sets in the two sequences $\{E^{(n)}\}$ and $\{F^{(n)}\}$? In the case $p = 2$, W.K. Hayman has shown that this is possible in the case of rarefied sets: he gave a talk on the subject at the BMO-seminar in Joensuu, Finland, Aug. 17–20, 1987. Using the general theory from [3] and [4], we shall in the present paper solve this problem for all dimensions $p \geq 2$ and for both types of exceptional sets (cf. Theorems 1 and 2).

§2 Notation

(i) We write D as the union of disjoint, half-open Whitney cubes $\{Q_k\}$ with sides parallel to the coordinate axes (cf. [4, Sections 4,5], [1, pp. 6–7], [6, p. 16]). If $Q \in \{Q_k\}$, we have

$$d(Q) = \text{dist.}\,(Q, \partial D) \geq 2\text{diam.}\,Q/\sqrt{p} \geq C(p)\,d(Q).$$

Let $\tilde{Q}$ be the double of Q, i.e. $\tilde{Q} = \{x \in \mathbb{R}^p : x - x_Q = 2(y - x_Q)$ for some $y \in Q\}$ where x_Q is the centre of Q. Then $\tilde{Q} \subset D$ and there is a constant B only depending on the dimension p such that $\tilde{Q}$ meets at most B cubes in the collection $\{\tilde{Q}_k\}$. To a Whitney cube $Q = Q_k$, we associate the numbers $(t, r, R) = (t_k, r_k, R_k)$: t is the distance from x_Q to ∂D, $2r$ is the side-length of the cube and $R = |x_Q|$ is the distance from the centre of Q to the origin. We shall also need $\theta = \arccos(t/R)$.

(ii) If $x = (x_1, \ldots, x_p)$, then $\tilde{x} = (-x_1, x_2, \ldots, x_p)$ is the reflection of x in ∂D.

(iii) For x and y in D, we define

$$G(x,y) = \begin{cases} |x - y|^{2-p} - |x - \tilde{y}|^{2-p}, & p \geq 3, \\ \log(|x - \tilde{y}|/|x - y|), & p = 2. \end{cases}$$

For $x \in D$ and $y \in \partial D$, we define $P(x,y) = x_1 |x - y|^{-p}$.

Let μ and μ_1 be nonnegative Radon measures with supports in $D \cup \partial D$ and ∂D, respectively. The Green potential and the Poisson integral of these measures are defined by

$$G\mu(x) = \int_D G(x,y)d\mu(y), \quad P\mu_1(x) = \int_{\partial D} P(x,y)d\mu_1(y), \quad x \in D.$$

(iv) If $y \in \partial D$, we define the cone $\Gamma(y) = \{x \in D : 2x_1 > |x - y|\}$.

(v) Let 0 be a relatively open subset of ∂D. The tent set $T(0)$ is defined
by

$$T(0) = D \setminus \left(\bigcup_{y \in \partial D \setminus 0} \Gamma(y) \right).$$

In the case $p = 2$, 0 is a disjoint union of open intervals $\cup I_j$, and the tent
set is simply a union of similar triangles with bases $\{I_j\}$. The triangles are
relatively closed in D.

(vi) The symbols C, C_p or A denote absolute positive constants (which
may depend on p), whose values are unimportant and may change from line
to line.

(vii) We shall say that two positive functions f and g are comparable,
written $f \approx g$, iff there exist constants A and C such that $Ag \le f \le Cg$.

(viii) Let S_1 be the class of all positive superharmonic functions u on D
for which there exist non-negative Radon measures μ_1 on ∂D and μ_2 on D
such that $u = P\mu_1 + G\mu_2$ with

$$\int_{\partial D} (1 + |y|)^{-p} d\mu_1(y) + \int_D (1 + |y|)^{-p} y_1 \, d\mu_2(y) < \infty,$$

(cf. [3, Definition 4.1]).

§3 Definitions of capacities

The ordinary capacity of a set $E \subset R^p$ is denoted by $c(E)$.

Given $E \subset D$, suppose that there exists a measure λ_E whose Green
potential is $G\lambda_E = \hat{R}^E_{x_1}$, where $\hat{R}^E_{x_1}$ is the regularized reduced function of
x_1 on E with respect to the cone of positive superharmonic functions on D.
We call λ_E the fundamental distribution on E and $\lambda_E(D) = \lambda(E)$ the outer
charge of E. We define $\gamma(E) = \int G\lambda_E(x) d\lambda_E(x)$ and call $\gamma(E)$ the Green
energy of E (cf. [5, p. 129], [3, p. 237]).

Given $E \subset D$, suppose there exist measures μ_1 and μ_2 on ∂D and D,
respectively, which are such that

$$\hat{R}^E_1(x) = \int_{\partial D} P(x, y) d\mu_1(y) + \int_D G(x, y) d\mu_2(y).$$

The Green mass of E is defined as $\lambda'(E) = \mu_1(\partial D) + \int_D y_1 d\mu_2(y)$ (cf. [3, p.
239]).

All these set functions are first defined for compact sets and then extended to capacities defined for general sets in a standard way (cf. [3,

p. 243]). They are all monotone and countably subadditive. Furthermore, we have $\lambda(E) = \lambda'(E)$ for any set $E \subset D$ (cf. [3, Lemma 2.5]).

Remark. The basic definitions and results on rarefied and minimally thin sets do not depend on the general assumption in [3] that the dimension p is at least 3: they are true also in the case $p = 2$.

§4   On Wiener conditions in terms of ordinary capacity

In Theorems 1 and 2, E is a subset of D and $E_k = E \cap Q_k$ for all Whitney cubes in our collection $\{Q_k\}$.

Theorem 1. a) *A sufficient condition for $E \subset D$ to be rarefied at infinity in D is that*

$$\sum \cos \theta_k (\log(3t_k/c(E_k)))^{-1} < \infty, \quad p = 2 \tag{4.1a}$$

$$\sum \cos \theta_k \, c(E_k) R_k^{2-p} < \infty, \qquad p \geq 3. \tag{4.1b}$$

b) *Let E be rarefied at infinity in D. Then there exists a relatively open set $0 \subset \partial D$ such that $\int_0 (1+|x|)^{1-p} dx' < \infty$ and such that (4.1a) or (4.1b) holds if we sum over all Whitney cubes which do not intersect $T(0)$. (In the integral above, we integrate with respect to Lebesgue measure on ∂D.)*

Remark. The union of the Whitney cubes which intersect $T(0)$ is rarefied at infinity.

Theorem 2. a) *A sufficient condition for $E \subset D$ to be minimally thin at infinity in D is that*

$$\sum (\cos \theta_k)^2 (\log(3t_k/c(E_k)))^{-1} < \infty, \quad p = 2, \tag{4.2a}$$

$$\sum (\cos \theta_k)^2 c(E_k) R_k^{2-p} < \infty, \qquad p \geq 3. \tag{4.2b}$$

b) *Let E be minimally thin at infinity in D. Then there exists a measure ν supported by ∂D such that the Poisson integral $P\nu$ is convergent and such that (4.2a) or (4.2b) holds if we sum over all Whitney cubes which do not intersect the set*

$$M(\nu) = \{x \in D : \int_{\partial D} |x - y|^{-p} d\nu(y) \geq 1\}$$

Remark. We note that $M(\nu) = \{x \in D : P\nu \geq x_1\}$. This set is minimally thin at infinity in D and it follows that

$$\int_M (1 + |x|)^{-p} dx < \infty,$$

where $M = M(\nu)$ (cf. [4, Corollary 3 p. 397] and the remark p. 398).

§5 Proofs of Theorems 1 and 2

In the proofs, we need several lemmas. Let Λ be a measure on $\{y \in \partial D : 1/2 \leq |y| \leq 4\}$ and let $\|\Lambda\|$ be the total mass of Λ. We shall also need the maximal function

$$N\Lambda(y) = \sup P\Lambda(x), \quad x \in \Gamma(y), \quad y \in \partial D.$$

Lemma 1. *Let $G = \{y \in \partial D : N\Lambda(y) > A\}$. Then G is an open set and*

$$|G| \leq C \|\Lambda\| / A, \tag{5.1}$$

where $|.|$ denotes Lebesgue measure on ∂D.

The proof is given in [4, Section 6].

Lemma 2. *Let Q be a Whitney cube in $D \subset R^2$ and let E be a subset of Q. Then we have*
$$\lambda(E) = \lambda'(E) \approx t(\log(4t/c(E)))^{-1}. \tag{5.2}$$

Proof. It suffices to prove the lemma assuming that E is a compact set of positive capacity. Since $G\lambda_E(x) = x_1$ p.p. on E, it follows that

$$\int_E \log |x - y| d\lambda_E(y) \leq \lambda(E) \log(4t) - t/2 \quad \text{p.p. on } E.$$

Using [2, Lemma 4] we deduce that

$$2\lambda(E) \geq t(\log(4t/c(E)))^{-1}.$$

To prove an inequality going the other way, we note that there is a unit measure μ supported by E such that

$$\log c(E) = \int_E \log |x - y| d\mu(x) \quad \text{p.p. on } E.$$

It follows that p.p. on E,

$$G\mu(x) = -\log c(E) + \int_E \log |x - \tilde{y}| d\mu(y) \geq \log(8t/5c(E)).$$

Integrating with respect to the fundamental distribution λ_E on E and using the fact that $E \subset Q$, we obtain

$$\lambda(E) \leq 2t(\log(2t/c(E)))^{-1} < 8t(\log(4t/c(E)))^{-1}.$$

This is clear since $c(E) \leq c(Q) \leq r\sqrt{2} \leq t$. We have proved Lemma 2. $\square$

Lemma 3. *Let Q be a Whitney cube in $D \subset R^p$, $p \geq 3$, and let E be a subset of Q. Then we have*

$$\lambda(E) = \lambda'(E) \approx t\, c(E), \quad \gamma(E) \approx t^2\, c(E). \tag{5.3}$$

Proof. Again, we assume that E is a compact set of positive capacity. We use the elementary inequality

$$|x - \tilde{y}|^{2-p} \leq C_p G(x, y) \leq 4C_p(p - 2)x_1 y_1 |x - y|^{2-p}|x - \tilde{y}|^{-2},$$

which is valid for $p \geq 3$.

Since $G\lambda_E(x) = x_1$ p.p. on E, the upper estimate of $G(x, y)$ implies that

$$Ct^2 \leq \int_E |x - y|^{2-p} y_1 d\lambda_E(y) \quad \text{p.p. on } E. \tag{5.4}$$

Let μ be a measure of total mass $c(E)$ supported by E such that $\int_E |x - y|^{2-p} d\mu(y) = 1$ p.p. on E. Integrating (5.4) with respect to μ, we obtain

$$Ct^2\, c(E) \leq \gamma(E) \leq 2t\, \lambda(E).$$

Conversely, it follows from the lower estimate of $G(x, y)$ that $G\mu(x) \geq (1 + C_p)^{-1}$ p.p. on E. Integrating with respect to λ_E, we obtain

$$2t(1 + C_p)c(E) \geq \lambda(E).$$

We have proved Lemma 3. $\square$

Let for a while E be a set contained in $D \cap \{x \in I\!R^p : 1 \leq |x| \leq 2\}$ and let $\{Q_k\}$ be a collection of Whitney cubes in D which intersect this half-annulus. If $E_k = E \cap Q_k$, we wish to compare $\lambda'(E)$ and $\sum \lambda'(E_k)$.

Lemma 4. *There are absolute constants A and C such that*

$$\sum{}' \lambda'(E_k) \leq C\lambda'(E), \tag{5.5}$$

where the prime in the summation means that we sum over all Whitney cubes intersecting $D \cap \{1 \leq |x| \leq 2\}$ except those which intersect also a tent set built on a relatively open subset of ∂D of measure of at most $A\lambda'(E)$.

Remark. It follows from the discussion in [4, Section 6] that the base of the tent is contained in $\partial D \cap \{1/2 \leq |x| \leq 4\}$.

Proof. To our given set E, there exist measures μ_1 and μ_2 such that $\hat{R}_1^E = P\mu_1 + G\mu_2$ and $\lambda'(E) = \mu_1(\partial D) + \int_D y_1 d\mu_2(y)$. Arguing as in [4, Section 5], we introduce

$$I(x) = \int_{\tilde{Q}_k} G(x,y)d\mu_2(y), \quad x \in Q_k,$$

$$J(x) = P\mu_1(x) + \int_{D \setminus \tilde{Q}_k} G(x,y)d\mu_2(y), \quad x \in Q_k.$$

The functions I and J are defined in the union of all Whitney cubes considered here and we have $\hat{R}_1^E(x) = I(x) + J(x)$, $x \in D$. From formula (6.8) in [4], we see that there is a nonnegative measure ν on ∂D such that

$$J(x) \leq CP\nu(x), \quad \nu(\partial D) \leq C_1 \left(\int_D x_1 d\mu_2(x) + \mu_1(\partial D) \right) = C_1 \lambda'(E).$$

From Lemma 1, we deduce that $J(x) \leq CP(x) \leq 1/2$ outside a tent set built on a relatively open set on ∂D of $(p-1)$-dimensional measure at most $C_2 \| \nu \| \leq A\lambda'(E)$. Let us discard all Whitney cubes which intersect this tent set. In the remaining cubes, we have $\hat{R}_1^E(x) \leq 2I(x)$, $x \in E$, and

$$\hat{R}_1^{E_k}(x) \leq 2I(x), \quad x \in Q_k \cap E.$$

Integrating with respect to the fundamental distribution $d\lambda_{E_k}$ on E_k, we obtain

$$\lambda(E_k) \leq 2 \int_{Q_k} x_1 d\mu_2(x),$$

$$\sum{}' \lambda'(E_k) = \sum{}' \lambda(E_k) \leq 2B \int x_1 d\mu_2(x) \leq 2B\lambda'(E),$$

since $\tilde{Q}$ meets at most $B-1$ cubes from $\{\tilde{Q}\}$. This concludes the proof of Lemma 4. $\square$

In our next lemma, we let E and $\{Q_k\}$ be as in the introduction to Lemma 4.

Lemma 5. *There are absolute constants A and C and a measure ν on ∂D with $\nu(\partial D) \leq A\gamma(E)$ such that*

$$\sum{}' \gamma(E_k) \leq C\gamma(E), \tag{5.6}$$

where the prime in the summation means that we sum over all Whitney cubes intersecting $D \cap \{1 \leq |x| \leq 2\}$ except those which intersect the set

$$\{x \in D : \int_{\partial D} |x-y|^{-p} d\nu(y) > 1\}.$$

Remark. It follows from the discussion in [4, Section 6] that the support of ν is contained in $\partial D \cap \{1/2 \leq |x| \leq 4\}$.

Proof. We argue in the same way as in the proof of Lemma 4. We write

$$\hat{R}^E_{x_1}(x) = \int G(x,y)d\mu(x) = I(x) + J(x),$$

where

$$I(x) = \int_{\tilde{Q}_k} G(x,y)d\mu(y), \quad x \in Q_k; \quad J(x) = G\mu(x) - I(x).$$

We note that $\gamma(E) = \int x_1 d\mu(x)$. Again, there exists a measure ν_0 on ∂D such that

$$J(x) < CP\nu_0(x), \quad \nu_0(D) \leq C_1 \int x_1 d\mu(x)$$

and we have $J(x) \leq C\,P\nu_0(x) \leq x_1/2$ (and thus $G\mu(x) \leq 2I(x)$) outside a set of the type described in the lemma. We omit the details. $\square$

Proof of Theorem 1. If $p \geq 3$, we use Lemma 3 and see that

$$\lambda'(E_k)R_k^{1-p} \approx \cos\theta_k\; c(E_k)R_k^{2-p}.$$

Let $\sum^{(n)}$ denote that we sum over all Whitney cubes which intersect the half-annulus $D \cap \{2^n \leq |x| < 2^{n+1}\}$. Since λ' is subadditive, we see that

$$\lambda'(E^{(n)}) \leq \sum^{(n)} \lambda'(E_k) \leq C \sum^{(n)} t_k\, c(E_k).$$

Multiplying by $2^{n(1-p)} \approx R_k^{1-p}$ and summing over n, we see that (1.2) holds if (4.1b) holds. Thus (4.1b) is a sufficient condition for E to be rarefied at infinity when $p > 3$. The details are similar in the case $p = 2$. We have proved Theorem 1a.

We give the details of the proof of Theorem 1b in the case $p \geq 3$. We start by applying Lemma 4 to the set $E^{(n)}/2^n$. This set is contained in $D \cap \{1 \leq |x| \leq 2\}$ and has the Green mass $\lambda'(E^{(n)})2^{-n(p-1)}$.

Restricting ourselves to those Whitney cubes which do not intersect the tent set $T(E^{(n)}/2^n)$ and summing over n, we see that the left hand member of (4.1b) is majorized by

$$C \sum_n 2^{n(1-p)} \sum^{(n)} \lambda'(E_k) \leq C_1 \sum 2^{n(1-p)}\lambda'(E^{(n)}) < \infty.$$

We have used Lemma 3, Lemma 4 and in the last step our assumption that E is rarefied at infinity.

Let n be fixed. Going back to the nth annulus, we see that the $(p-1)$-dimensional measure of the base of the associated tent set is at most $A\lambda'(E^{(n)})$. Let 0 be the union of all these sets. Since E is rarefied at infinity in D, it follows from (1.2) that $\int_0 (1+|x|^{1-p})dx'$ is convergent. This completes the proof of Theorem 1.

In the proof of Theorem 2, we use Lemma 5 instead of Lemma 4. In this case, assuming that $p \geq 3$, the Green energy of the set $E^{(n)}/2^n$ is $2^{-np}\gamma(E^{(n)})$. Thus, if (4.2b) holds, it follows from Lemma 3 that (1.1) will hold, i.e., the set E is minimally thin at infinity in D.

Conversely, if E is minimally thin at infinity, it is clear that (4.2b) holds provided that we sum over Whitney cubes which do not intersect a certain set near ∂D. According to Lemma 5, the set $E^{(n)}$ gives us a measure ν_n on ∂D with $\nu_n(\partial D) \leq A\gamma(E^{(n)})$. Let $\nu = \sum_1^\infty \nu_n$. From (1.1), we see that $\int_{\partial D}(1+|y|)^{-p}d\nu(y)$ is convergent. Furthermore, the set to be avoided by the Whitney cubes in the summation in (4.2b) is contained in the set $M(\nu)$ mentioned in Theorem 2.

The details are similar when $p = 2$. This concludes the proof of Theorem 2. $\square$

References

[1] L. Carleson, Selected problems on exceptional sets. Van Nostrand 1967.

[2] M. Essén, W.K. Hayman and A. Huber, Slowly growing subharmonic functions I. Comment. Math. Helv. *52* (1977), 329–356.

[3] M. Essén and H.L. Jackson, On the covering properties of certain exceptional sets in a half-space, Hiroshima Math. J. *10* (1980), 233–262.

[4] M. Essén, H.L. Jackson and P.J. Rippon, On minimally thin and rarefied sets in $I\!\!R^p$, $p \geq 2$, Hiroshima Math. J. *15* (1985), 393–410.

[5] J. Lelong-Ferrand, Etude au voisinage de la frontière des fonctions surharmoniques positives dans un demi-espace. Ann. Sci. Ecole Norm. Sup (3) *66*, (1949), 125–159.

[6] E.M. Stein, Singular integrals and differentiability properties of functions, Princeton University Press 1970.

Department of Mathematics
University of Uppsala
Thunbergsvägen 3
S-752 38 Uppsala
Sweden

F.W. Gehring and G.J. Martin

The Matrix and Chordal Norms of Möbius Transformations

§1 Introduction

A family F of self homeomorphisms of the Möbius space $\bar{R}^n$ is said to have the *convergence property* if each infinite subfamily of F contains a sequence $\{f_j\}$ such that one of the following is true.

A. There exists a self homeomorphism f of $\bar{R}^n$ such that $f_j \to f$ and $f_j^{-1} \to f^{-1}$ uniformly in $\bar{R}^n$.

B. There exist points x_0, y_0 in $\bar{R}^n$ such that $f_j \to y_0$ and $f_j^{-1} \to x_0$ locally uniformly in $\bar{R}^n \setminus \{x_0\}$ and $\bar{R}^n \setminus \{y_0\}$, respectively.

A family G of self homeomorphisms of $\bar{R}^n$ is said to be a *convergence group* if it has the convergence property and forms a group under composition.

The notion of a convergence group appears to capture the essential compactness properties of Möbius transformations required to establish the elementary theory of discrete Möbius groups in the plane and higher dimensions. See [GM1], [H], [MS], [MT], [T].

We let $q(x, y)$ denote the *chordal distance* between $x, y \in \bar{R}^n$, that is

$$q(x, y) = |p(x) - p(y)|,$$

where p denotes stereographic projection of $\bar{R}^n$ onto the unit sphere S^n in R^{n+1}. We say that $x, y \in \bar{R}^n$ are *antipodal points* if $q(x, y) = 2$, that is, if $p(x)$ and $p(y)$ are diametrically opposite points of S^n.

Next we introduce the *chordal metric*

$$d(f, g) = \sup(q(f(x), g(x)) : x \in \bar{R}^n)$$

on the space of self homeomorphisms of $\bar{R}^n$ and call

$$d(f) = d(f, id)$$

the *chordal norm* of f. Then $d(f)$ measures the maximum chordal deviation of f from the identity, $0 \le d(f) \le 2$ and $d(f) = 2$ if and only if f maps one point of a pair of antipodal points of $\bar{R}^n$ onto the other. Note also that

$$d(f, g) = d(f \circ g^{-1}, id) = d(f \circ g^{-1}).$$

If f is a Möbius transformation acting on the extended complex plane $\bar{R}^2$, then f has a standard representation by a matrix A in $SL(2, \mathbb{C})$ which is unique up to a factor -1. It is customary to call $\| A \|$ the *matrix norm* of f, where for any matrix $C = \begin{pmatrix} a & b \\ c & d \end{pmatrix}$ in $GL(2, \mathbb{C})$, $\| C \|$ denotes the euclidean norm

$$\| C \|^2 = |a|^2 + |b|^2 + |c|^2 + |d|^2.$$

In this case, $\sqrt{2} \le \| A \| < \infty$ and $\| A \| = \sqrt{2}$ if and only if f is a chordal isometry. An alternative matrix norm which better measures the deviation of f from the identity is

$$\min(\| A - I \|, \| A + I \|).$$

The purpose of this note is to point out a discreteness criterion for convergence groups in terms of the chordal norm and to establish a pair of sharp inequalities which relate the above chordal and matrix norms for Möbius transformations acting on $\bar{R}^2$. Such inequalities yield lower bounds for the chordal norm in a nonelementary discrete group of Möbius transformations and thus express in a very geometric way the isolation of elements in such a group [GM3].

§2   Discreteness

A group G of homeomorphisms acting on $\bar{R}^n$ is said to be *discrete* if it contains no sequence of distinct elements which converge to a homeomorphism f uniformly in $\bar{R}^n$. Alternatively, G is discrete if there exists a constant $r > 0$ such that $d(f, g) \ge r$ for each pair of distinct elements $f, g \in G$.

The following is an analogue for convergence groups of a familiar discreteness criterion for Möbius groups.

2.1 Theorem. *If G is a convergence group, then the following are equivalent.*

A. G is discrete.

B. $\operatorname{card}\{f \in G : d(f) \le r\} < \infty$ for some $r \in (0,2)$.

C. $\operatorname{card}\{f \in G : d(f) \le r\} < \infty$ for each $r \in (0,2)$.

Proof. Suppose that G is discrete, fix $r \in (0,2)$ and let

$$G_0 = \{f \in G : d(f) \le r\}. \tag{2.2}$$

If G_0 is not finite, then by the convergence property, there exists a sequence of distinct $f_j \in G_0$ and points $x_0, y_0 \in \bar{I\!R}^n$ such that

$$f_j \to y_0 \quad \text{locally uniformly in} \quad \bar{I\!R}^n \setminus \{x_0\}. \tag{2.3}$$

Since $r < 2$, we can choose $z_0 \in \bar{I\!R}^n \setminus \{x_0\}$ so that $q(z_0, y_0) > r$. Then

$$q(f_j(z_0), y_0) \ge q(z_0, y_0) - q(f_j(z_0), z_0) \ge q(z_0, y_0) - r > 0$$

for all j contradicting (2.3). Thus G_0 is finite and A implies C.

Suppose next that B holds for some $r \in (0,2)$, let G_0 be as in (2.2) and set

$$r_0 = \min\{d(f) : f \in G_0 \setminus \{id\}\} > 0.$$

Then for each pair of distinct elements $f, g \in G$, $h = f \circ g^{-1} \ne id$,

$$d(f,g) = d(h) \ge r_0 > 0$$

and hence G is discrete. $\square$

2.4 Remark. Suppose that G is a convergence group acting on $\bar{I\!R}^n$. Then Theorem 2.1 implies that G is discrete if and only if, except for a finite number of elements, each $g \in G$ maps some point $x = x_g$ onto a point y which is almost antipodal to x.

§3 Upper bound for $\|A\|$

We establish next the following relation between the norms $d(f)$ and $\| A \|$.

3.1 Theorem. *Suppose that f is a Möbius transformation acting on $\bar{I\!R}^2$. Then*

$$\| A \|^2 \le 2\frac{4 + d(f)^2}{4 - d(f)^2} \le \frac{4}{2 - d(f)}. \tag{3.2}$$

This inequality is sharp for each value of $d(f)$.

Theorem 3.1 is an immediate consequence of the following result which is of independent interest.

3.3 Theorem. *Suppose that f is a Möbius transformation acting on $\bar{R}^n$ and that $\tilde{f}$ is its Poincaré extension to H^{n+1}. Then*

$$\rho(\tilde{f}(e_{n+1}), e_{n+1}) \leq \log\left[\frac{2 + d(f)}{2 - d(f)}\right], \qquad (3.4)$$

where ρ denotes the hyperbolic metric in H^{n+1} with curvature -1 and $e_{n+1} = (0, \ldots, 0, 1)$. This inequality is sharp for each value of $d(f)$.

Proof of Theorem 3.3. We may assume that $\rho(\tilde{f}(e_{n+1}), e_{n+1}) > 0$ since otherwise there is nothing to prove.

Let $\tilde{p}$ denote the Möbius transformation which maps $\bar{H}^{n+1}$ onto the closed unit ball $\bar{B}^{n+1}$ and whose restriction to $\bar{R}^n$ is stereographic projection. Then $h = \tilde{p} \circ \tilde{f} \circ \tilde{p}^{-1}$ is a Möbius self map of $\bar{B}^{n+1}$,

$$0 < \rho(\tilde{f}(e_{n+1}), e_{n+1}) = \log\left[\frac{1 + |h(0)|}{1 - |h(0)|}\right] \qquad (3.5)$$

and

$$|h(y) - y| = |p \circ f(x) - p(x)| = q(f(x), x) \leq d(f) \qquad (3.6)$$

for $y \in S^n$, where $x = p^{-1}(y)$. Then $h(0) \neq 0$ by (3.5) and hence $h(\infty) \neq \infty$.

Let L denote the line through 0 and $h^{-1}(\infty)$. Then L meets S^n in diametrically opposite points z and $-z$, $h(L)$ is a line through $h(0)$ which is orthogonal to S^n and hence $h(z) = -h(-z)$. By relabeling we may arrange that

$$|h(0) - h(-z)| < |h(0) - h(z)|. \qquad (3.7)$$

By (3.5) we can choose $r \in (0, 2)$ so that

$$\frac{2 + r}{2 - r} = \frac{1 + |h(0)|}{1 - |h(0)|}. \qquad (3.8)$$

Let Σ denote the n-dimensional spherical cap

$$\Sigma = \left\{ y \in S^n : |y - z|^2 \leq 2 - r \right\},$$

fix $y \in S^n \setminus \Sigma$ and choose $s \in (-2, 2]$ so that

$$|h(y) - h(z)|^2 = 2 + s.$$

Then

$$|y + z|^2 = 4 - |y - z|^2 < 2 + r, \qquad |h(y) + h(z)|^2 = 4 - |h(y) - h(z)|^2 = 2 - s.$$

By the invariance of the absolute cross ratio under h and (3.7),

$$\frac{|y + z|}{|y - z|} = \frac{|h(y) - h(-z)|}{|h(y) - h(z)|} \frac{|h(0) - h(z)|}{|h(0) - h(-z)|} = \frac{|h(y) + h(z)|}{|h(y) - h(z)|} \frac{1 + |h(0)|}{1 - |h(0)|}.$$

If $s = 2$, then $s > r$; if $s < 2$, then

$$\frac{\sqrt{2 + r}}{\sqrt{2 - r}} \frac{\sqrt{2 + s}}{\sqrt{2 - s}} > \frac{|y + z|}{|y - z|} \frac{|h(y) - h(z)|}{|h(y) + h(z)|} = \frac{1 + |h(0)|}{1 - |h(0)|} = \frac{2 + r}{2 - r}$$

by (3.8) and again $s > r$. Thus

$$|h(y) - h(z)|^2 > 2 + r$$

and we conclude that

$$\{y \in S^n : |y - h(z)|^2 \le 2 + r\} \subseteq h(\Sigma). \tag{3.9}$$

Next if $y \in \Sigma$, then

$$|h(y) - z|^2 = |h(y) - y + y - z|^2 = -2((h(y) - y) \cdot z) + |y - z|^2$$
$$\le 2|h(y) - y| + |y - z|^2 \le 2q(f(x), x) + 2 - r$$
$$\le 2 + 2d(f) - r$$

where $x = p^{-1}(y)$, and

$$h(\Sigma) \subseteq \{y \in S^n : |y - z|^2 \le 2 + 2d(f) - r\}. \tag{3.10}$$

Now (3.9) and (3.10) imply that $d(f) \ge r$; inequality (3.4) then follows from (3.5) and (3.8).

To see that (3.4) is sharp, fix $a \in (1, \infty)$ and set $f(x) = a^2 x$. Then

$$\rho(\tilde{f}(e_{n+1}), e_{n+1}) = \log(a^2) \tag{3.11}$$

while for each $x = \frac{y}{a} \in \bar{R}^n$,

$$q(f(x), x)^2 = \frac{4|ay - \frac{y}{a}|^2}{(|ay|^2 + 1)(|\frac{y}{a}|^2 + 1)} = \frac{4(a - a^{-1})^2}{|y|^2 + a^2 + a^{-2} + |y|^{-2}}$$
$$\le 4\frac{(a - a^{-1})^2}{(a + a^{-1})^2}$$

with equality when $|y| = 1$. Hence

$$d(f) = 2\frac{a^2 - 1}{a^2 + 1},\tag{3.12}$$

and we obtain equality in (3.4). $\square$

Proof of Theorem 3.1 Let $n = 2$ in Theorem 3.3. Then

$$\| A \|^2 = 2\cosh(\rho(\tilde{f}(e_3), e_3))$$

by [B, Theorem 4.2.1] and (3.2) follows directly from (3.4). $\square$

§4 Upper bound for $\|A-I\|$

Though inequality (3.2) is sharp for each value of $d(f)$, it does not yield much information when $d(f)$ is small. For example, when $d(f) = 0$, it implies only that f is a chordal isometry instead of that $f = id$. We derive here a different version for (3.2) with the norm $\| A \|$ replaced by $\min(\| A - I \|, \| A + I \|)$.

4.1 Theorem. *Suppose that f is a Möbius transformation acting on $\bar{\mathbb{R}}^2$. Then there exists a matrix A in $SL(2, \mathbb{C})$ which represents f such that*

$$\| A - I \|^2 \leq \frac{4d(f)^2}{4 - d(f)^2}.\tag{4.2}$$

This inequality is asymptotically sharp as $d(f) \to 2$.

Proof. Suppose first that f has two fixed points and multiplier a^2. By conjugating by a preliminary chordal isometry, we may assume that $\mathrm{fix}(f) = \{-r, r\}$, where $r \in (0, 1]$, and hence that f is represented by the matrix

$$A = \frac{1}{2}\begin{bmatrix} a + a^{-1} & r(a - a^{-1}) \\ r^{-1}(a - a^{-1}) & a + a^{-1} \end{bmatrix}.$$

By replacing A by $-A$ we may further assume that $\mathrm{Re}(a) \geq 0$ and hence that

$$t = \left|\frac{a - 1}{a + 1}\right| \leq 1.\tag{4.3}$$

Then

$$4 \| A - I \|^2 = |a - a^{-1}|^2(2t^2 + r^2 + r^{-2}).\tag{4.4}$$

Next from [GM2] it follows that

$$\frac{4}{d(f)} = \begin{cases} \dfrac{r}{t} + \dfrac{t}{r} & \text{if } r > t, \\ 2 & \text{if } r \leq t. \end{cases} \tag{4.5}$$

Since (4.2) is trivially true when $d(f) = 2$, we need only consider the case where $r \in (t, 1]$. Then (4.4) and (4.5) imply that

$$16 \parallel A - I \parallel^2 (4d(f)^{-2} - 1) = |a - a^{-1}|^2 g(r, t), \tag{4.6}$$

where

$$g(r, t) = \left(2t^2 + r^2 + r^{-2}\right)\left(\frac{r}{t} - \frac{t}{r}\right)^2 .$$

Now for each fixed $t \in (0, 1)$, $g(r, t)$ is increasing in r for $r \in (t, 1]$. Hence

$$g(r, t) \leq g(1, t) = 2(t^2 + 1)(t^{-1} - t)^2 \leq 4\left(\left|\frac{a+1}{a-1}\right| - \left|\frac{a-1}{a+1}\right|\right)^2 \tag{4.7}$$

$$\leq 64|a - a^{-1}|^{-2}$$

and (4.2) follows from (4.6) and (4.7).

If f has a single fixed point, then by conjugating by a chordal isometry we may assume that f is represented by

$$A = \begin{bmatrix} 1 & 2a \\ 0 & 1 \end{bmatrix}$$

and hence that

$$\parallel A - I \parallel^2 = 4|a|^2. \tag{4.8}$$

In this case

$$\frac{4}{d(f)} = \begin{cases} |a| + |a|^{-1} & \text{if } |a| < 1, \\ 2 & \text{if } |a| \geq 1 \end{cases} \tag{4.9}$$

by [GM2]. Thus we may assume that $|a| < 1$, and we obtain

$$\parallel A - I \parallel^2 (4d(f)^{-2} - 1) = (|a|^2 - 1)^2 < 1,$$

and hence (4.2), from (4.8) and (4.9).

To show that (4.2) is asymptotically sharp, set $f(x) = a^2 x$ where $a \in (1, \infty)$. Then

$$\parallel A - I \parallel^2 = (1 + a^{-2})(a - 1)^2$$

while

$$\frac{4d(f)^2}{4 - d(f)^2} = (1 + a^{-1})^2(a - 1)^2$$

by (3.12). Thus

$$\| A - I \|^2 \left[\frac{4d(f)^2}{4 - d(f)^2} \right]^{-1} \to 1$$

as $a \to \infty$ and hence as $d(f) \to 2$. $\square$

§5 Final remarks

In some cases, it is better to replace

$$\min(\| A - I \|, \| A + I \|) \quad \text{by} \quad \| A - A^{-1} \|$$

when measuring how far the Möbius transformation f represented by A differs from the identity. For example, the quantity $\| A - A^{-1} \|$ does not depend on the choice of the two possible matrices in $SL(2, \mathbb{C})$ which represent f. For this norm we have the following analogue of Theorem 4.1.

5.1 Theorem. *Suppose that f is a Möbius transformation acting on $\bar{\mathbb{R}}^2$ and that A is a matrix in $SL(2, \mathbb{C})$ which represents f. Then*

$$\| A - A^{-1} \|^2 \leq \frac{8d(f)^2}{4 - d(f)^2}. \tag{5.2}$$

This inequality is sharp for each value of $d(f)$.

Equality holds in (5.2) whenever f is hyperbolic with antipodal fixed points. Sharper versions can be established for the case where f is elliptic or parabolic. For all of the above results, see [GM2].

References

[B] A.F. Beardon, *The geometry of discrete groups*, Springer-Verlag 1983.

[GM1] F.W. Gehring and G.J. Martin, Discrete quasiconformal groups I, *Proc. London Math. Soc.* (3) 55 (1987) 331–358.

[GM2] F.W. Gehring and G.J. Martin, Inequalities for matrices and Möbius transformations (to appear).

[GM3] F.W. Gehring and G.J. Martin, Discreteness in Kleinian groups and the iteration theory of quadratic mappings (to appear).

[H] A. Hinkkanen, Abelian and nondiscrete convergence groups on the circle (to appear).

[MS] G.J. Martin and R. Skora, Group actions on the 2-sphere, *Amer. J. Math.* (to appear).

[MT] G.J. Martin and P. Tukia, Convergence and Möbius groups, Holomorphic Functions and Moduli II, Math. Sci. Res. Inst. Publ. 11 (1988) 113–140.

[T] P. Tukia, Homeomorphic conjugates of Fuchsian groups, *J. Reine Angew. Math.* (to appear).

The research of both authors was supported in part by the National Science Foundation, Grants DMS-87-02356 and DMS-86-02550.

University of Michigan
Ann Arbor, Michigan

Yale University
New Haven, Connecticut

W. K. Hayman, Ch. Pommerenke

On Meromorphic Functions with Growth Conditions

§1 Functions of locally bounded characteristic

We assume that the function f is meromorphic in $D = \{z \in C : |z| < 1\}$. Let $C_1, C_2, \ldots$ denote positive constants. We say that f has locally bounded characteristic (l.b.c) in D if there exists a function φ analytic and univalent in D with $\varphi(D) \subset D$, $\varphi(D \cap R) \subset R$ and

$$\varphi(1) = 1, \quad \varphi'(1) < \infty, \tag{1.1}$$

such that

$$T_o(r, f(e^{i\Theta}\varphi(w))) \leq C_1 \quad (0 \leq \Theta \leq 2\pi). \tag{1.2}$$

Here $\varphi(1)$ is the angular limit and $\varphi'(1)$ the angular derivative; by the Julia-Wolff lemma [9, p.306], condition (1.1) is equivalent to

$$1 - \varphi(w) < C_2(1 - w) \quad \text{for} \quad 0 < w < 1. \tag{1.3}$$

Furthermore T_o denotes the Ahlfors-Shimizu characteristic [6, p.12].

This class of functions was introduced by B. Korenblum and the first author [7] where the Nevanlinna characteristic T was used instead of T_o. If f is analytic this affects only the constants because [6, p.13]

$$|T(r, g) - T_o(r, g) - \log^+ |g(0)|| \leq \frac{1}{2} \log 2.$$

In the present case of a meromorphic function the Ahlfors-Shimizu characteristic turns out to be more natural because it is invariant under spherical rotations of f.

As an example, we state a special case of a theorem of P.J. Rippon.

Theorem A [12, Theorem 2]. *If f has l.b.c. then*

$$\frac{|1 - ze^{-i\Theta}|^2}{1 - |z|^2} \log |f(z)| \to \alpha(\Theta) \tag{1.4}$$

as $z \to e^{i\Theta}$ along almost all rays from $\boldsymbol{D}$ ending at $e^{i\Theta}$. Furthermore $\alpha(\Theta)$ is bounded and is zero outside a countable set.

The following result of B. Korenblum and the first author is the starting point of the present paper.

Theorem B [7, Theorem 5]. *If f is analytic in $\boldsymbol{D}$ and if*

$$|f(re^{i\Theta})| \leq e^{k(r)} \quad \text{for} \quad re^{i\Theta} \in \boldsymbol{D}, \tag{1.5}$$

where $k(r)$ is positive, continuous and nondecreasing in $[0,1)$ and satisfies

$$\int_0^1 \left(\frac{k(r)}{1-r} \right)^{1/2} dr < \infty, \tag{1.6}$$

then f has l.b.c. in $\boldsymbol{D}$.

Our first theorem is a generalization to meromorphic functions. The statements will be invariant under spherical rotations, in particular replacing f by $1/f$.

We assume throughout that $\lambda(r)$ is positive and nondecreasing and that $\delta(r)$ is nonincreasing and satisfies $0 < \delta(r) < 1$. For $0 \leq r < 1$ and $z_0 \in \boldsymbol{D}$ we define

$$\Delta(z_0, r) = \left\{ z \in \boldsymbol{D} : \left| \frac{z - z_0}{1 - \bar{z}_0 z} \right| \leq \delta(r) \right\}. \tag{1.7}$$

By a spherical disk, we mean a halfplane or the inner or outer domain of a circle.

Theorem 1. *Suppose that f is meromorphic in $\boldsymbol{D}$ and that, for $|z_0| \leq r < 1$,*

$$f(\Delta(z_0, r)) \text{ omits a spherical disk of chordal radius } e^{-\lambda(r)}. \tag{1.8}$$

Then f has l.b.c in $\boldsymbol{D}$ if

$$\int_0^1 \left(\frac{\lambda(r)}{1-r} \right)^{1/2} \frac{1}{\delta(r)} dr < \infty. \tag{1.9}$$

Suppose now that (1.5) holds. Then $f(z)$ omits $\{|w| > e^{k(r)}\}$ if $|z| \leqq r$, and this is a spherical disk of centre ∞ and chordal radius less than $e^{-k(r)}$. Since

$$z \in \Delta(z_0, r) \quad \text{implies} \quad |z| \leqq \frac{r + \delta(r)}{1 + r\delta(r)} = r', \qquad (1.10)$$

we see that our assumptions are satisfied with $\delta(r) = 1/2$ and $\lambda(r) = k(r')$ where $r' = (r + 1/2)(1 + r/2)$. Hence Theorem B is a consequence of Theorem 1.

In the case that f has no poles, we can prove a stronger (though less invariant) result.

Theorem 2. *Suppose that f is analytic in* $\boldsymbol{D}$ *and that, for $|z_1| \leqq r < 1$, $|z_2| \leqq r$,*

$$|f(z_1)| \leqq e^{-\lambda(r)}, \quad |f(z_2)| = e^{\lambda(r)} \quad imply \quad \left| \frac{z_2 - z_1}{1 - \bar{z}_1 z_2} \right| = \delta(r). \qquad (1.11)$$

Then, for $|z_1| \leq r < 1$, $|z_2| \leq r$,

$$|f(z_1)| \leqq e^{-\lambda^*(r)}, \quad |f(z_2)| = e^{\lambda^*(r)} \quad imply \quad \left| \frac{z_2 - z_1}{1 - \bar{z}_1 z_2} \right| = \frac{1}{6} \qquad (1.12)$$

where

$$\lambda^*(r) = 8\lambda\left(\frac{1+r}{2}\right) \bigg/ \delta\left(\frac{1+r}{2}\right), \qquad (1.13)$$

and f has l.b.c. if

$$\int_0^1 \left(\frac{\lambda(r)}{(1-r)\delta(r)}\right)^{1/2} dr < \infty. \qquad (1.14)$$

We write

$$f^{\#}(z) = \frac{|f'(z)|}{1 + |f(z)|^2} \qquad (1.15)$$

for the spherical derivative and define

$$\mu(r) = \max_{|z| \leqq r} f^{\#}(z) \quad \text{for} \quad 0 \leq r < 1. \qquad (1.16)$$

The function f is said to be normal if

$$\mu(r) = \frac{O(1)}{1 - r} \quad \text{as} \quad r \to 1. \qquad (1.17)$$

Normal functions need not have bounded characteristis, but do have l.b.c. For the latter conclusion, (1.17) can be significantly weakened.

Theorem 3. *If f is meromorphic in $\boldsymbol{D}$ and if*

$$\int_0^1 (1 - r)^{1/2} \mu(r) dr < \infty, \tag{1.18}$$

then f has l.b.c in $\boldsymbol{D}$. If f is analytic then (1.18) can be replaced by

$$\int_0^1 \mu(r)^{1/2} dr < \infty. \tag{1.19}$$

§2 Integral representations

An important aspect of functions with locally bounded characteristic is the connection with integral representations.

Theorem C [7, Theorem 6]. *Suppose that h is analytic in $\boldsymbol{D}$ with $h(0) = 0$ such that*
$$f(z) = \exp[zh'(z)] \quad (z \in \boldsymbol{D}) \tag{2.1}$$

has l.b.c. in $\boldsymbol{D}$. Then

$$\Psi(t) = \lim_{r \to 1} \operatorname{Im} h(re^{it}) \quad (0 \le t \le 2\pi) \tag{2.2}$$

exists. The function Ψ is bounded and we have

$$h(z) = \frac{iz}{\pi} \int_0^{2\pi} \frac{\Psi(t)}{e^{it} - z} dt \quad (z \in \boldsymbol{D}). \tag{2.3}$$

This is an adaptation of a result of B. Korenblum and the first author. The present function h is connected with the function g of [7] by $g(z) = zh'(z)$. We use h in order to stress the relation to functions of bounded mean oscillation.

Theorem D [7, Theorem 5][14]. *Suppose that h is analytic in $\mathbb{D}$ with $h(0) = 0$ and that*

$$\mathrm{Re}[zh'(z)] \le k(r) \quad for \quad |z| = r < 1, \tag{2.4}$$

where (see (1.6))

$$\int_0^1 \left(\frac{k(r)}{1-r} \right)^{1/2} dr < \infty. \tag{2.5}$$

Then the real-valued function Ψ defined by (2.2) is bounded, the one-sided limits $\Psi(t\pm)$ exist for all t, and the integral representation (2.3) holds.

This is an immediate consequence of Theorems B and C except for the fact that Ψ has only jump discontinuities which is due to K. Samotij [14]; he has given a direct proof of Theorem D that also generalizes to harmonic functions in higher dimensions. See also the survey article [13] of P.J. Rippon.

The function h has a representation of the form (2.3) with bounded real-valued Ψ if and only if

$$\sup_{z \in \mathbb{D}} |\mathrm{Im}\, h(z)| < \infty. \tag{2.6}$$

We shall consider four spaces of functions analytic in $\mathbb{D}$. By definition, $h \in \mathcal{B}_0$ if

$$(1 - |z|^2)|h'(z)| \to 0 \quad as \quad |z| \to 1 \tag{2.7}$$

while $h \in \mathcal{B}$ ("Bloch") if $(1 - |z|^2)|h'(z)|$ is bounded in $\mathbb{D}$. Furthermore, $f \in \mathrm{VMOA}$ ("vanishing mean oscillation") if

$$\int_{|\zeta|=1} \frac{1 - |z|^2}{|\zeta - z|^2} |h'(\zeta)|^2 |d\zeta| \to 0 \quad as \quad |z| \to 1 \tag{2.8}$$

while $f \in \mathrm{BMOA}$ ("bounded mean oscillation") if the integral is bounded in $\mathbb{D}$; see e.g. [2][5, Chapter 5]. There are the following (strictly one-sided) inclusions:

$$\{h : |\mathrm{Im}\, h| \text{ bounded}\} \subset \mathrm{BMOA} \subset \mathcal{B}, \tag{2.9}$$

$$\{h : \mathrm{Im}\, h \text{ continuous in} \bar{\mathbb{D}}\} \subset \mathrm{VMOA} \subset \mathcal{B}_0. \tag{2.10}$$

The following theorem shows that (2.5) is the critical growth rate; compare also [8].

Theorem E [7, Theorem 4]. *Suppose that the integral (2.5) diverges. Then there is a function h that satisfies (2.4), while $h \notin \mathcal{B}$.*

Now we turn to the question of continuity.

Theorem 4. *Suppose that h satisfies (2.4) and (2.5). If $h \in \mathcal{B}_o$ then $\mathrm{Im}\, h$ is continuous in $\bar{\mathbf{D}}$ and Ψ is continuous in $[0, 2\pi]$.*

Thus we see that, if we restrict ourselves to functions satisfying (2.4) and (2.5), equality holds in (2.10). An example of a function, not in $\mathcal{B}_o$, but satisfying (2.4) and (2.5), is $\log(1 - z)$.

Finally we turn to integral representations (2.3) with complex-valued Ψ.

Theorem 5. *Suppose that h is analytic in $\mathbf{D}$ with $h(0) = 0$ and that*

$$\mathrm{Re}[zh'(z)] \leq k(r) \quad for \quad |z| = r < 1, \tag{2.11}$$

where now

$$\int_0^1 k(r)dr < \infty. \tag{2.12}$$

(i) *If $h \in \mathcal{B}$ then $f \in BMOA$, and (2.3) holds for some bounded complex function Ψ.*

(ii) *If $h \in \mathcal{B}_o$ then $f \in VMOA$, and (2.3) holds for some continuous complex function Ψ.*

Since (2.12) is weaker than (2.5) we cannot delete our assumption $h \in \mathcal{B}$ as Theorem E shows. We will discuss these questions further in the final section.

§3 Proof of Theorem 1

(a) We construct the conformal mapping φ that occurs in the definition that f has l.b.c. We define

$$r_n = 1 - 2^{-n}, \ t_n = c(1 - r_n)^{1/2}\delta(r_n)\lambda(r_n)^{-1/2}, \tag{3.1}$$

where c is chosen such that $t_1 = 1$, and furthermore

$$\rho(t) = 1 - 2^{-n} \quad \text{for} \quad t_{n+1} \leq t < t_n, \quad n = 1, 2, \ldots. \tag{3.2}$$

We note that

$$\left.\begin{array}{c}\displaystyle\int_0^1 \frac{1-\rho(t)}{t^2}\,dt = \sum_{n=1}^{\infty} 2^{-n} \int_{t_{n+1}}^{t_n} t^{-2}\,dt < \sum_{n=1}^{\infty} 2^{-n} t_{n+1}^{-1} \\[4mm] \displaystyle = 2c^{-1} \sum_{n=2}^{\infty} (1-r_n)^{1/2} \lambda(r_n)^{1/2} \delta(r_n)^{-1} < \infty\end{array}\right\} \tag{3.3}$$

by our assumption (1.9) and by (3.1).

Let φ map $\boldsymbol{D}$ conformally onto the Jordan domain

$$G = \{re^{it} : 0 < r < \rho(|t|), -1 < t < 1\} \tag{3.4}$$

in $\boldsymbol{D}$ such that $\varphi(w)$ is real for $0 < w < 1$ and $\varphi(1) = 1$. It follows from (3.3) by a theorem of Warschawski [16] (see e.g. [15, p.366]) that $\varphi'(1) < \infty$. To complete the proof we have to show that (1.2) holds for some constant C_1 independent of Θ. We may assume that $\Theta = 0$; the constants $C_3, \ldots$ below will not change if we replace $f(z)$ by $f(e^{i\Theta}z)$.

(b) We suppose that $z_0 \in \boldsymbol{D}$ and that $|z_0| < r < 1$. We write

$$\Delta_1(z_0, \rho) = \left\{ z \in \boldsymbol{D} : \left| \frac{z - z_0}{1 - \bar{z}_0 z} \right| \leq \rho \right\} \quad (0 < \rho < 1) \tag{3.5}$$

so that $\Delta(z_0, r) = \Delta_1(z_0, \delta(r))$ by (1.7).

We now define

$$A(z_0, \rho) = \frac{1}{\pi} \iint_{\Delta_1(z_0,\rho)} f^{\#}(z)^2\,dx\,dy \tag{3.6}$$

and claim that

$$A(z_0, \delta(r)/2) \leq \lambda(r)/\log 2. \tag{3.7}$$

To prove (3.7) we note that, by (1.8), the set $f(\Delta(z_0, r))$ omits a disk of chordal radius at least $\exp(-\lambda(r))$. Since $f^{\#}$ is invariant under rotations of the sphere, we may assume that ∞ is the centre of the omitted disk. Hence

$$(1 + |f(z)|^2)^{1/2} \leq e^{\lambda(r)} \quad \text{for} \quad z \in \Delta(z_0, r).$$

We now define

$$f_o(z) = f((z_0 + z)/(1 + \bar{z}_0 z)) \quad (z \in \boldsymbol{D})$$

so that f_0 is analytic and

$$1 + |f(z)|^2 \leq e^{2\lambda(r)} \quad \text{for} \quad |z| < \delta(r). \tag{3.8}$$

We also write

$$A_o(\rho) = \frac{1}{\pi} \iint_{|z|<\rho} f_o^{\#}(z)^2 dxdy, \tag{3.9}$$

so that, by (3.6),

$$A_o(\rho) = A(z_0, \rho). \tag{3.10}$$

We deduce [6, p.12] from (3.9) and (3.8) that

$$\int_o^{\delta(r)} A_o(\rho)\rho^{-1}d\rho \leq \frac{1}{4\pi} \int_o^{2\pi} \log(1 + |f_o(\delta(re^{it}))|^2)dt \leq \lambda(r)$$

and therefore that

$$A_o\left(\frac{\delta(r)}{2}\right) \log 2 \leq \int_{\delta(r)/2}^{\delta(r)} A_o(\rho)\rho^{-1}d\rho \leq \lambda(r).$$

This proves (3.7) because of (3.10).

(c) We write

$$g(w) = f(\varphi(w)) \quad (w \in \boldsymbol{D}) \tag{3.11}$$

and

$$B(t) = \frac{1}{\pi} \iint_{|w|<t} g^{\#}(w)^2 dudv \quad (0 < t < 1). \tag{3.12}$$

It follows from Schwarz's lemma that

$$|\varphi(w)| \leq r \equiv (t + \varphi(0))/(1 + \varphi(0)t) \quad \text{for} \quad |w| \leq t. \tag{3.13}$$

Hence we see from (3.11) that $g^{\#}(w) \leq C_3$ for $|w| \leq 1/2$ and thus [6, p.12]

$$T_o(1, g) = \int_0^1 \frac{B(t)}{t} dt < C_4 + 2 \int_{1/2}^1 B(t)dt. \tag{3.14}$$

Since φ is univalent it follows from (3.11), (3.12) and (3.13) that

$$B(t) \leq \frac{1}{\pi} \iint\limits_{G \cap \{|z| \leq r\}} f^{\#}(z)^2 \, dx dy, \tag{3.15}$$

where r is given by (3.13). Thus (3.14) implies

$$T_o(1, g) < C_4 + C_5 \int\limits_{1/2}^{1} \left(\iint\limits_{G \cap \{|z| \leqq r\}} f^{\#}(z)^2 \, dx dy \right) dr. \tag{3.16}$$

For $n = 2, 3, \ldots$ we define

$$\delta_n = 2^{-n-2} \delta(r_n), \quad G_n = \{ r_{n-1} \leqq |z| < r_n \} \cap G. \tag{3.17}$$

It follows from the definition (3.4) of G and from (3.2) that G_n can be covered by disks

$$\{ |z - z_\nu| < \delta_n \} \quad (\nu = 1, \ldots, N_n) \quad \text{with} \quad |z_\nu| \leqq r_n.$$

By (1.9) and (3.1), $\delta_n = O(r_n - r_{n-1})$ and $\delta_n = O(t_n)$. Thus we can choose N_n such that

$$N_n \leqq C_6 2^{-n} t_n / \delta_n^2 \leqq C_7 2^{n/2} \lambda(r_n)^{-1/2} \delta(r_n)^{-1} \tag{3.18}$$

because of (3.1). If $|z - z_\nu| < \delta_n$ then

$$\left| \frac{z - z_\nu}{1 - \bar{z}_\nu z} \right| = \left| \frac{z - z_\nu}{1 - |z_\nu|^2 + \bar{z}_\nu(z_\nu - z)} \right| < \frac{\delta_n}{2^{-n} - \delta_n} < 2^{n+1} \delta_n$$

by (3.17). It follows that

$$G_n \subset \bigcup_{\nu=1}^{N_n} \Delta_1(z_\nu, 2^{n+1} \delta_n)$$

and therefore from (3.17) and (3.7) that

$$\iint\limits_{G_n} f^{\#}(z)^2 \, dx dy \leqq N_n \lambda(r_n) / \log 2.$$

Hence we conclude from (3.16), (3.17) and (3.18) that

$$T_o(1, g) < C_4 + C_5 \sum_{n=2}^{\infty} \int_{r_{n-1}}^{r_n} \left(\iint_{G \cap \{|z| < r_n\}} f^{\#}(z)^2 \, dx \, dy \right) dr$$

$$\leqq C_4 + C_5 \sum_{n=2}^{\infty} 2^{-n} \left(\sum_{m=2}^{n} \iint_{G_m} f^{\#}(z)^2 \, dx \, dy + C_8 \right)$$

$$= C_9 + 2C_5 \sum_{m=2}^{\infty} 2^{-m} \iint_{G_m} f^{\#}(z)^2 \, dx \, dy$$

$$< C_9 + C_{10} \sum_{m=2}^{\infty} 2^{-m} N_m \lambda(r_m)$$

$$< C_9 + C_{11} \sum_{m=2}^{\infty} 2^{-m/2} \lambda(r_m)^{1/2} \delta(r_m)^{-1} = C_{12} < \infty$$

by (3.3).

§4   Proof of Theorem 2

We deal first with the case that λ and δ are constants.

Lemma 1. *Suppose that f is analytic in $\boldsymbol{D}$ and that (1.11) holds where λ and δ are constants. Then, for $|z_1| \leqq r < 1$, $|z_2| \leqq r$,*

$$|f(z_1)| \leqq e^{-8\lambda/\delta}, \ |f(z_2)| = e^{8\lambda/\delta} \quad imply \quad \left| \frac{z_2 - z_1}{1 - \bar{z}_1 z_2} \right| = \frac{1}{2}. \qquad (4.1)$$

Proof. Suppose that

$$|f(0)| = e^{a}, \quad a = 8\lambda/\delta. \qquad (4.2)$$

If $|f(z)| = e^{-\lambda}$ in $\boldsymbol{D}$ there is nothing to prove. Suppose then that r_1 and r_2 are the largest numbers such that

$$|f(z)| > e^{-\lambda} \quad \text{for} \quad |z| < r_1; \quad |f(z)| > e^{\lambda} \quad \text{for} \quad |z| < r_2.$$

Then there exists $z_1 = r_1 e^{i\theta}$ such that $|f(z_1)| = e^{-\lambda}$. Also by our construction, if $z_2 = r_2 e^{i\theta}$, we have $|f(z_2)| = e^{\lambda}$. Hence we deduce from (1.11) that

$$\frac{r_1 - r_2}{1 - r_1 r_2} = \left| \frac{z_1 - z_2}{1 - \bar{z}_1 z_2} \right| = \delta. \tag{4.3}$$

On the other hand, if

$$\varphi(z) = e^{\lambda} f(r_1 z) \quad (z \in \boldsymbol{D})$$

then $|\varphi(z)| > 1$, and now classical estimates yield

$$\log |\varphi(z)| = \frac{1 - |z|}{1 + |z|} \log |\varphi(0)| \quad \text{for} \quad z \in \boldsymbol{D}.$$

We may choose z such that $|r_1 z| = r_2$ and $|\varphi(z)| = e^{2\lambda}$. This yields, by (4.2),

$$2\lambda = (a + \lambda)(r_1 - r_2)/(r_1 + r_2).$$

Using also (4.3) we obtain

$$4\lambda = \lambda(r_1 + 3r_2) = a(r_1 - r_2) = a\delta(1 - r_1 r_2) = a\delta(1 - r_2).$$

Thus if $a = 8\lambda/\delta$, we have $r_2 = 1/2$. We deduce that

$$|f(0)| = e^{8\lambda/\delta} \quad \text{implies} \quad |f(z)| > e^{\lambda} \quad (|z| < 1/2). \tag{4.4}$$

Now we apply this result, for given $z_2 \in \boldsymbol{D}$, to

$$g(z) = f\left(\frac{z_2 + z}{1 + \bar{z}_2 z} \right) \quad (z \in \boldsymbol{D}).$$

Clearly g satisfies the analogue of (1.11) also. Hence it follows from (4.4) that

$$|f(z_2)| = e^{8\lambda/\delta} \quad \text{implies} \quad |f(z_1)| > e^{\lambda} \quad \text{if} \quad \left| \frac{z_1 - z_2}{1 - \bar{z}_2 z_1} \right| < \frac{1}{2}.$$

Thus, (4.1) is satisfied and Lemma 1 is proved.

We now consider the general case of Theorem 2. We choose a fixed value r with $0 < r < 1$ and write

$$R = (1 + r)/2, \quad \lambda_1 = \lambda(R), \quad \delta_1 = \delta(R), \quad a = 8\lambda_1/\delta_1.$$

We proceed to show that, for $|z_1| \leqq r$, $|z_2| \leqq r$,

$$|f(z_1)| \leqq e^{-a}, \quad |f(z_2)| = e^a \quad \text{imply} \quad \left| \frac{z_2 - z_1}{1 - \bar{z}_1 z_2} \right| = \frac{1}{6}. \qquad (4.5)$$

This will prove assertion (1.12) of Theorem 2.

It follows from the maximum principle (applied to the variable z_2) that

$$\varphi_R(z_1, z_2) \equiv \left| \frac{R(z_1 - z_2)}{1 - R^2 \bar{z}_1 z_2} \right| \leqq \varphi_1(z_1, z_2) \equiv \left| \frac{z_1 - z_2}{1 - \bar{z}_1 z_2} \right| \qquad (4.6)$$

for $|z_1| \leq 1$ and $|z_2| \leq 1$. Consider now $g(z) = f(Rz)$. By hypothesis, if $|z_1| \leqq R$, $|z_2| \leqq R$ then

$$|f(z_1)| \leqq e^{-\lambda_1}, \quad |f(z_2)| = e^{\lambda_1} \quad \text{imply} \quad \varphi_1(z_1, z_2) \geq \delta_1.$$

Thus, writing Rz_1 and Rz_2 instead of z_1, z_2, we deduce that

$$|g(z_1)| \leqq e^{-\lambda_1}, \quad |g(z_2)| = e^{\lambda_1} \quad \text{imply} \quad \varphi_R(z_1, z_2) = \delta_1.$$

Using (4.6) we deduce that $\varphi(z_1, z_2) = \delta_1$. Hence Lemma 1 shows that

$$|g(z_1)| \leqq e^{-a}, \quad |g(z_2)| = e^a \quad \text{imply} \quad \varphi(z_1, z_2) = 1/2.$$

We apply this conclusion with z_1/R, z_2/R instead of z_1, z_2 and deduce that

$$|z_1| < R, \quad |z_2| < R, \quad |f(z_1)| \leqq e^{-a}, \quad |f(z_2)| = e^a$$

imply

$$\left| \frac{z_2/R - z_1/R}{1 - \bar{z}_1 z_2/R^2} \right| > \frac{1}{2}. \qquad (4.7)$$

To prove (4.5) we show that

$$\left| \frac{z_2 - z_1}{1 - \bar{z}_1 z_2} \right| \Big/ \left| \frac{z_2/R - z_1/R}{1 - \bar{z}_1 z_2/R^2} \right| > \frac{1}{3} \quad \text{for} \quad |z_1| \leqq r, \ |z_2| \leqq R. \qquad (4.8)$$

By the maximum principle we may assume that $|z_1| = r$ and $|z_2| = R$. Then the left-hand side of (4.8) is

$$\left| \frac{z_2 - z_1}{1 - \bar{z}_1 z_2} \right| = \frac{R - r}{1 - rR} = \frac{1}{2 + r} > \frac{1}{3}.$$

In particular (4.8) holds if $|z_1| \leqq r$ and $|z_2| \leqq r$, so that (4.7) implies

$$\left| \frac{z_2 - z_1}{1 - \bar{z}_1 z_2} \right| > \frac{1}{6}$$

which yields (4.5).

Finally we assume that (1.14) holds and put

$$\delta_1(r) = \frac{1}{12}, \quad \lambda_1(r) = \lambda^* \left(\frac{1+r}{2} \right) = 8\lambda \left(\frac{3+r}{4} \right) \Big/ \delta \left(\frac{3+r}{4} \right).$$

Then

$$\int_0^1 \left(\frac{\lambda_1(r)}{1 - r} \right)^{1/2} \frac{1}{\delta_1(r)} dr = 24\sqrt{8} \int_{3/4}^1 \left(\frac{\lambda(t)}{(1 - t)\delta(t)} \right)^{1/2} dt < \infty$$

by (1.14). Also if $|z_0| \leqq r$ and $z_1, z_2 \in \Delta(z_0, r)$, we have

$$|z_j| \leqq \frac{1+r}{2} \quad (j = 1, 2), \quad \left| \frac{z_2 - z_1}{1 - \bar{z}_1 z_2} \right| \leqq \frac{1/12 + 1/12}{1 + (1/12)^2} < \frac{1}{6}.$$

Thus we cannot have $|f(z_1)| < e^{-\lambda_1(r)}$ and $|f(z_2)| > e^{-\lambda_1(r)}$. Hence one or other of these two inequalities is false throughout $\Delta(z_0, r)$. Therefore f has l.b.c. by Theorem 1, and Theorem 2 is proved.

§5 Proof of Theorem 3

Suppose that $|z_0| \leqq r$ and define $\delta(r)$ by

$$\delta(r)^{-1} = 6[1 + (1 - r)\mu((1 + r)/2)].$$

Then if $z_1 \in \Delta(z_0, r)$ and $\zeta = (z_1 - z_0)/(1 - \bar{z}_0 z_1)$, we have $|\zeta| < \delta(r) < 1/6$. Thus

$$|z_1 - z_0| = \frac{(1 - |z_0|^2)|\zeta|}{|1 + \bar{z}_0 \zeta|} < \frac{12(1 - r)\delta(r)}{5} < \frac{1 - r}{2},$$

so that $|z_1| \leqq (1+r)/2$. With $\mu(r)$ given by (1.16), we deduce that

$$\left| \tan^{-1} |f(z_1)| - \tan^{-1} |f(z_0)| \right| = \int\limits_{[z_0, z_1]} f^\#(z)|dz|$$

$$\leqq \frac{12}{5} \delta(r)(1-r)\mu\left(\frac{1+r}{2}\right) < \frac{2}{5}.$$

Hence, if $|f(z_0)| \leqq 1$, then

$$\tan^{-1} |f(z_1)| < \frac{\pi}{4} + \frac{2}{5} < \tan^{-1} e, \quad \text{thus} \quad |f(z_1)| < e.$$

Similarly if $|f(z_0)| = 1$, we deduce that $|f(z_1)| > e^{-1}$. Thus in either case f satisfies the hypotheses of Theorem 1 with $\lambda(r) = 1$. Also

$$\int\limits_0^1 \left(\frac{\lambda(r)}{1-r}\right)^{1/2} \frac{1}{\delta(r)} dr = 6 \int\limits_0^1 (1-r)^{-1/2} \left[1 + (1-r)\mu\left(\frac{1+r}{2}\right)\right] dr < \infty$$

if (1.18) holds, so that in this case f has l.b.c.

If f is analytic we apply Theorem 2 and deduce that f has l.b.c. if

$$\int\limits_0^1 (1-r)^{-1/2} \delta(r)^{-1/2} dr < \infty$$

which is true if (1.19) holds. This completes the proof of Theorem 3.

§6 Proof of Theorem 4

By Theorem D, the function Ψ defined by (2.2) has one-sided limits. We claim that $\Psi(\theta-) = \Psi(\theta+)$ for all θ. It suffices to consider $\theta = 0$. We define

$$a = \Psi(0-) - \Psi(0+), \quad \Psi_0(t) = \begin{cases} 0 & \text{for } -\pi < t \leqq 0 \\ a & \text{for } 0 < t \leqq \pi \end{cases}$$

and obtain from (2.3) that

$$h'(z) = \frac{i}{\pi} \int\limits_0^{2\pi} \frac{e^{it}\Psi(t)}{(e^{it} - z)^2} dt = -\frac{2a}{\pi(1-z^2)} + \frac{i}{\pi} \int\limits_{-\pi}^{\pi} \frac{e^{it}(\Psi + \Psi_0)}{(e^{it} - z)^2} dt.$$

It follows that, as $r \to 1-$,

$$\left|(1-r)^2 h'(r) + \frac{2a}{\pi}\right| \leq \frac{1}{\pi} \int\limits_{-\pi}^{\pi} (1-r^2)\frac{|\Psi + \Psi_0|}{|e^{it} - r|^2} dt \to 0$$

because $\Psi + \Psi_0$ is continuous at 0. Hence $h \in \mathcal{B}_0$ implies $a = 0$, by (2.7). We conclude that $\Psi(\theta-) = \Psi(\theta+)$ so that Ψ is continuous.

Furthermore (2.2) shows that $\int_{-\pi}^{\pi} \Psi(t)dt = 0$. Hence we see from (2.3) that

$$\operatorname{Im} h(z) = \frac{1}{2\pi} \int\limits_{-\pi}^{\pi} \frac{1 - |z|^2}{|e^{it} - z|^2} \Psi(t)dt \quad (z \in \boldsymbol{D}), \tag{6.1}$$

and since Ψ is continuous, $\operatorname{Im} h$ has a continuous extension to $\bar{\boldsymbol{D}}$.

§7 Functions of bounded mean oscillation

The function h belongs to BMOA if and only if both

(a) $h \in H^1$

(b) $h(e^{it}) \in \text{BMO}(\boldsymbol{R})$;
   it is possible [2, p.16] to replace condition (b) by

(b′) $\operatorname{Re} h(e^{it}) \in \text{BMO}(\boldsymbol{R})$.

Similar characterizations hold for VMOA and VMO.

Lemma 2. *If h is analytic in $\boldsymbol{D}$ and satisfies (2.11) and (2.12), then $h \in H^1$ and furthermore*

$$\iint\limits_{\boldsymbol{D}} |h'(z)|dxdy < \infty. \tag{7.1}$$

The class of functions that satisfy (7.1) form a space that is (equivalent to the) dual of $\mathcal{B}_0$ and whose dual is $\mathcal{B}$; see [4] or [1, Theorem 2.4].

Proof. Suppose that $0 < r < 1$ and write $\rho = (1+r)/2$ and $g(z) = zh'(z)$. Since $\operatorname{Re}[k(\rho) - g(\rho\zeta)] > 0$ for $\zeta \in \boldsymbol{D}$ by (2.11), it follows that

$$\rho|g'(\rho\zeta)| \leq 2\operatorname{Re}[k(\rho) - g(\rho\zeta)]/(1 - |\zeta|^2) \quad \text{for} \quad |\zeta| < 1.$$

If we integrate this inequality and use that $g(0) = 0$, we obtain

$$\int_0^{2\pi} |g'(re^{it})|dt \leq \frac{4\pi k(\rho)/\rho}{1 - r^2/\rho^2} \leq \frac{8\pi k((1+r)/2)}{1-r}.$$

Integrating over $r \in [0, s]$, we deduce that

$$s\int_0^{2\pi} |h'(se^{it})|dt = \int_0^{2\pi} |g(se^{it})|dt \leq 8\pi \int_0^s k\left(\frac{1+r}{2}\right)\frac{dr}{1-r}$$

and therefore

$$\iint_{I\!D} |h'(z)|dxdy \leq 8\pi \int_0^1 \int_0^s k\left(\frac{1+r}{2}\right)\frac{dr}{1-r}ds = 8\pi \int_0^1 k\left(\frac{1+r}{2}\right)dr$$

which is finite by (2.12). Another integration then shows that $h \in H^1$.

Proof of Theorem 5. It is not difficult to see that (i) follows from [10, Satz 2] while the proof of (ii) is similar to that of [11, Theorem 2]. We restrict ourselves to the more complicated proof of (ii).

We assume that h satisfies (2.11) and (2.12) and belongs to $\mathcal{B}_0$. It follows from Lemma 2 that $h \in H^1$. Fix $z = re^{i\Theta}$ with $0 \leq r < 1$, $0 \leq \Theta \leq 2\pi$ and define

$$q(t) = \mathrm{Re}[h(e^{i\Theta + it}) - h(re^{i\Theta})], \quad p(t) = \frac{1}{2\pi}\frac{1 - r^2}{|e^{it} - r|^2}. \tag{7.2}$$

We write $I = [-\pi, \pi]$ and $A = \{t \in I : q(t) \geq 0\}$.

It follows from the Poisson integral formula that

$$\int_I pqdt = 0. \tag{7.3}$$

Since $e^x - 1 \geq x + \frac{x^2}{2}$ $(x \geq 0)$ and $e^x - 1 \geq x$ $(x \leq 0)$ we deduce from (7.3) that

$$\int_I (e^q - 1)pdt \geq \frac{1}{2}\int_A q^2pdt.$$

Hence we see from (7.3) and Schwarz's inequality that

$$\left(\int_I |q| p\,dt \right)^2 = \left(2 \int_A qp\,dt \right)^2 \leq 4 \int_A q^2 p\,dt \\ \leq 8 \int_I (e^q - 1)p\,dt. \tag{7.4}$$

Let $C_1, C_2, \ldots$ denote positive constants. It follows from (2.7) by integration that there exists a positive α_0 such that

$$|h(\rho e^{i\Theta}) - h(r e^{i\Theta})| \leq \frac{1}{2} \log \frac{1-\rho}{1-r} \quad \text{for} \quad 1 - \frac{\pi}{\alpha_0} \leq \rho \leq r. \tag{7.5}$$

We write $r_t = 1 - t/\alpha_0$ and deduce from (2.7) that

$$\left| h(r_t e^{i\Theta + it}) - h(r_t e^{i\Theta}) \right| \leq C_1 \tag{7.6}$$

and furthermore from (2.11) and (2.12) that

$$\operatorname{Re}[h(e^{i\tau}) - h(r_t e^{i\tau})] = \int_{r_t}^1 \operatorname{Re}[e^{i\tau} h'(s e^{i\tau})]ds \\ \leq \int_{r_t}^1 \frac{k(s)}{s} ds \leq C_2. \tag{7.7}$$

Hence it follows from (7.2), (7.5), (7.6) and (7.7) that, for $\alpha_0(1-r) \leq t \leq \pi$,

$$q(t) = \operatorname{Re}[h(e^{i\Theta + it}) - h(r_t e^{i\Theta + it}) + h(r_t e^{i\Theta + it}) - h(r_t e^{i\Theta})] \\ + \operatorname{Re}[h(r_t e^{i\Theta}) - h(r e^{i\Theta})] \\ \leq C_2 + C_1 + \frac{1}{2} \log \frac{t}{\alpha_0(1-r)} = C_3 + \frac{1}{2} \log \frac{t}{1-r}.$$

Since $p(t) \leq C_4(1-r)t^{-2}$ we deduce that, for $\alpha > \alpha_0$,

$$\int_{\alpha(1-r)}^{\pi} e^q p\,dt \leq C_5 \int_{\alpha(1-r)}^{\pi} \left(\frac{t}{1-r} \right)^{1/2} (1-r)t^{-2}dt \leq C_6 \alpha^{-1/2}.$$

If $\varepsilon > 0$ is given we can therefore choose α such that $\alpha \geq \alpha_0$ and

$$\int\limits_{\alpha(1-r)}^{\pi} (e^q - 1)pdt + \int\limits_{-\pi}^{-\alpha(1-r)} (e^q - 1)pdt < \varepsilon; \tag{7.8}$$

the second integral is handled in an analogous fashion.

As in (7.6) and (7.7), we obtain from (2.7), (2.11) and (2.12) that

$$\left|h(re^{i\Theta+it}) - h(re^{i\Theta})\right| < \varepsilon, \quad \mathrm{Re}[h(e^{i\tau}) - h(re^{i\tau})] < \varepsilon \tag{7.9}$$

if $r_0(\varepsilon) < r < 1$ and $|t| \leq \alpha(1 - r)$. This implies $q(t) < 2\varepsilon$ and therefore

$$\int\limits_{-\alpha(1-r)}^{\alpha(1-r)} (e^q - 1)pdt \leq (e^{2\varepsilon} - 1) \int\limits_{-\alpha(1-r)}^{\alpha(1-r)} pdt < e^{2\varepsilon} - 1.$$

Thus we conclude from (7.4) and (7.8) that, for $r_0(\varepsilon) \leq r < 1$,

$$\left(\int\limits_{-\pi}^{\pi} |q|pdt \right)^2 < 8\varepsilon + 8(e^{2\varepsilon} - 1)$$

and it follows from (7.2) that $\mathrm{Re}\, h(e^{it}) \in \mathrm{VMO}$; see [5. Theorem VI.5.1(c)]. Hence $h \in \mathrm{VMOA}$.

J.G. Clunie and P.J. Rippon [3] have shown that if h satisfies the two-sided condition

$$|\mathrm{Re}[zh'(z)]| \leq k(|z|), \quad \int\limits_0^1 k(r)dr < \infty$$

then h has a representation (2.3) with Ψ given by (2.2).

We give now an example to show that we cannot choose the bounded function Ψ in (2.3) to be real-valued if we only assume that the function $h \in \mathcal{B}$ satisfies our one-sided condition (2.11) with (2.12).

Suppose that $1/2 < b < 1$ and define h by $h(0) = 0$ and

$$zh'(z) = \frac{1}{1-z} \left(-1 + i \left(\log \frac{e}{1-z} \right)^{-b} \right) + 1 - i \tag{7.10}$$

for $z \in \boldsymbol{D}$. If we write $1 - z = e^{-s-it}$ then $s > -\log 2$ and $|t| < \pi/2$, and (7.10) yields

$$\mathrm{Re}[zh'(z)] = -e^s \cos t - e^s s^{-b} \sin t + O(e^s s^{-b-1}) \qquad (7.11)$$

as $s \to +\infty$, $|t| < \pi/2$. This expression is negative if $|t| \le \pi/2 - 2s^{-b}$ and s is large. On the other hand, if $\pi/2 - 2s^{-b} < |t| \le \pi/2$ and $|z| \le r$, then

$$1 - r \le 1 - |z|^2 = 2e^{-s} \cos t - e^{-2s} < 4e^{-s} s^{-b}$$

and thus, by (7.11), for large s,

$$\mathrm{Re}[zh'(z)] < 2e^s s^{-b} \le \frac{C_7}{1-r} \left(\log \frac{1}{1-r} \right)^{-2b}.$$

Hence we see that (2.11) holds with

$$k(r) = \frac{C_8}{1-r} \left(\log \frac{1}{1-r} \right)^{-2b}$$

so that (2.12) is satisfied because $b > 1/2$. But

$$\mathrm{Im}\, h(r) = \int_0^r \left[\frac{1}{1-x} \left(\log \frac{e}{1-x} \right)^{-b} - 1 \right] \frac{dx}{x} \to +\infty$$

as $r \to 1$ because $b < 1$. Hence (2.6) is false so that (2.3) cannot hold with a bounded real-valued function Ψ.

References

[1] J.M. Anderson, J. Clunie and Ch. Pommerenke, On Bloch functions and normal functions, J. reine angew. Math. 270 (1974), 12–37.

[2] A. Baernstein II, Analytic functions of bounded mean oscillation, Aspects of Contemporary Complex Analysis, Academic Press, London 1980, p.3–36.

[3] J.G. Clunie and P.J. Rippon, On harmonic functions with integrable maximum modulus, Ann. Acad. Sci. Fenn., Ser. A.I. Math. 8 (1983), 333–341.

[4] P.L. Duren, B.W. Romberg and A.L. Shields, Linear functionals on H^p spaces with $0 < p < 1$, J. reine angew. Math. 238 (1969), 32–60.

[5] J.B. Garnett, Bounded analytic functions, Academic Press, New York 1981.

[6] W.K. Hayman, Meromorphic functions, Oxford University Press 1975.

[7] W.K. Hayman and B. Korenblum, An extension of the Riesz-Herglotz formula, Ann. Acad. Sci. Fenn. Ser. A.I.2 (1976), 175–201.

[8] W.K. Hayman and B. Korenblum, A critical growth rate for functions regular in the disk, Michigan Math. J. 27 (1980), 21–30.

[9] Ch. Pommerenke, Univalent functions, Vandenhoeck & Ruprecht, Göttingen 1975.

[10] Ch. Pommerenke, Schlichte Funktionen and analytische Funktionen von beschränkter mittlerer Oszillation, Comment. Math. Helv. 52 (1977), 591–602.

[11] Ch. Pommerenke, On univalent functions, Bloch functions and VMOA, Math. Ann 236 (1978), 199–208.

[12] P.J. Rippon, The fine boundary behaviour of certain delta-subharmonic functions, J. London Math. Soc. (2) 26 (1982), 487–503.

[13] P.J. Rippon, Subharmonic functions with unilateral growth conditions, preprint 1986.

[14] K. Samotij, A representation theorem for harmonic functions in the ball in $I\!R^n$, Ann. Acad. Sci. Fenn. Ser. A.I. Math. 11 (1986), 29–37.

[15] M. Tsuji, Potential theory in modern function theory, Chelsea Publ. Corp., New York 1975.

[16] S.E. Warschawski, Über das Randverhalten der Ableitung der Abbildungsfunktion bei konformer Abbildung, Math. Z. 35 (1932), 321–456.

Department of Mathematics
University of York
York, YO1 5DD
England

Fachbereich Mathematik
Technische Universität
D-1000 Berlin 12
W. Germany

Maurice Heins

A Theorem of Wolff-Denjoy Type

§1 Introduction

The classical theorem concerning the iteration of analytic functions mapping $\Delta = \{|z| < 1\}$ into itself is due to Wolff [8, 9] and Denjoy [2]. It states that if such a function f does not have a fixed point in Δ, then $(f^{[n]})_0^\infty$, the sequence of iterates of f, tends to a point ζ of the unit circumference and, in addition,

$$\text{Re}\left[\frac{\zeta + f(z)}{\zeta - f(z)}\right] \geq \text{Re}\left(\frac{\zeta + z}{\zeta - z}\right), \quad |z| < 1. \tag{1.1}$$

The case of equality occurs in (1.1) for an allowed f if and only if f is a parabolic conformal automorphism of Δ with "fixed point" ζ. We recall that $(f^{[n]})_0^\infty$ is defined recursively by the requirements that $f^{[0]}$ be the identity map and $f^{[n+1]} = f \circ f^{[n]}$ for each nonnegative integer n.

Further, if (1.1) holds for some analytic function f mapping Δ into itself and ζ on the unit circumference, then either f is the identity map or $(f^{[n]})_0^\infty$ tends to ζ.

It is to be observed that the harmonic function on Δ given by $\text{Re}[(\zeta + z)/(\zeta - z)]$ is a minimal positive harmonic function on Δ normalized to take the value 1 at 0.

It is natural to ask to what extent the results of Wolff and Denjoy remain valid when Δ is replaced by a Riemann surface S admitting nonconstant positive harmonic functions and f by a holomorphic map of S into itself. The general problem does not appear to be simple. However, when S is restricted to belong to a "reasonable" class of such Riemann surfaces, a theorem of Wolff-Denjoy type holds. [For results without a complete counterpart of (1.1) for the case of plane multiply-connected regions, cf. [3].] The object of the present paper is to show that this is indeed the case when S has finite topological characteristics, i.e., genus g and number of boundary components c, satisfying $c \geq 1$, $2g + c \geq 2$, and no boundary component is pointlike.

We suppose, as we may, that S is a region of a compact Riemann surface T, the boundary Γ of S consisting of c disjoint regular analytic closed Jordan curves. When $g = 0$ and $c = 2$, we take S to be an annulus $\{r < |z| < r^{-1}\}$, $0 < r < 1$. Otherwise we assume, as we may, that T is the Schottky double of S. We fix a point $a \in S$ and restrict attention to the minimal positive harmonic functions on S normalized to take the value 1 at a. They are in $(1,1)$ correspondence with the points of Γ. Given $q \in \Gamma$, we denote the normalized minimal positive harmonic function on S vanishing continuously at the points of $\Gamma - \{q\}$ by u_q.

The following theorem of Wolff-Denjoy type will be established.

Theorem 1. *Let f denote a holomorphic map of S into itself. Then one and only one of the following statements holds:* (i) $(f^{[n]})_0^\infty$ *tends to a point* $\alpha \in S$, *and in this case* α *is the unique fixed point of f.* (ii) f *is a conformal automorphism of S.* (iii) $(f^{[n]})_0^\infty$ *tends to a point $q \in \Gamma$ and in this case*

$$u_q \circ f > u_q. \tag{1.2}$$

A map f satisfies

$$u_q \circ f = u_q \tag{1.3}$$

if and only if f is the identity map. If f satisfies (1.2), *then* $(f^{[n]})_0^\infty$ *tends to q.*

We remark that when $2g + c > 2$, in (ii) the conformal automorphism has finite order.

In the proof of Theorem 1 use will be made of a general iteration theorem for holomorphic maps of Riemann surfaces into themselves (§2), uniformization methods, the Lindelöf principle, the principle of hyperbolic length, and a generalization [4] of the "Starrheitssatz" of Aumann and Carathéodory [1].

I am informed by Dr. Alan Beardon that he has obtained theorems of Wolff-Denjoy type in a differential-geometric setting for contractive maps. Of course, there remains to be made for the situation of the present paper a comparison of the results achieved.

§2

We shall want to use the following iteration theorem (Theorem 2) for a holomorphic map f of a noncompact Riemann surface F into itself, the fundamental group of F being nonabelian. [The situation for compact Riemann surfaces of genus ≥ 2 is classical: if the map is not constant, it is a conformal automorphism; the group of conformal automorphisms is finite.] Theorem 2 is stated without proof in [5]. A proof will be given in this section.

Theorem 2. *Let f be a holomorphic map of a noncompact Riemann surface F with nonabelian fundamental group into itself. Exactly one of the following statements holds:* (i) *There exists a unique point $\alpha \in F$ such that $(f^{[n]})_0^\infty$ tends to α.* (ii) *f is a conformal automorphism of F of finite order.* (iii) *$(f^{[n]})_0^\infty$ tends to a unique Kerékjártó-Stoilow boundary element of F in the sense that given K_1 compact $\subset F$ there exists a unique component C of $F - K_1$ such that for each compact set $K_2 \subset F$ we have $f^{[n]}(K_2) \subset C$ for n large.*

In case (i) *the point α is the unique fixed point of f.*

The following proof finds its roots in Valiron's treatment of the iteration problem for a half-plane (equivalently, Δ) based on the theory of normal families, cf. [6, 7]. Here we use liftings of the $f^{[n]} \circ \varphi$ with respect to φ where φ is a fixed conformal universal covering of F with domain Δ. The disjunction of the statements (i), (ii), (iii) is immediate as is the last statement of the theorem.

We distinguish the following possibilities: I. There exists $p_0 \in F$ such that $(f^{[n]}(p_0))_0^\infty$ does not tend to the ideal boundary. II. The statement I does not hold, i.e., for each $p \in F$ the orbit $(f^{[n]}(p))_0^\infty$ tends to the ideal boundary of F.

I. We suppose that φ has been so normalized that $\varphi(0) = p_0$. There exists a subsequence $(f^{[\nu(n)]}(p_0))_0^\infty$ of $(f^{[n]}(p_0))_0^\infty$ which tends to a point of F. There exists r, $0 < r < 1$, such that all the points $f^{[\nu(n)]}(p_0)$ have preimages with respect to φ in $\{|z| \leq r\}$. We let g_n denote an analytic function mapping Δ into itself and satisfying $\varphi \circ g_n = f^{[\nu(n)]} \circ \varphi$ as well as the condition that $g_n(0)$ is a preimage of $f^{[\nu(n)]}(p_0)$ in $\{|z| \leq r\}$. We assume, as we may by taking a subsequence if necessary, that $(g_n)_0^\infty$ converges to an analytic function mapping Δ into itself. We introduce concurrently the analytic function h_n mapping Δ into itself satisfying the conditions $f^{[\nu(n+1)-\nu(n)]} \circ \varphi = \varphi \circ h_n$ and $h_n[g_n(0)] = g_{n+1}(0)$. As a consequence of these conditions we obtain the identity $g_{n+1} = h_n \circ g_n$.

Two cases may occur: (1) $\lim g_n$ is a constant, say taking the value b, (2) $\lim g_n$ is not a constant.

(1) From the definition of g_n we see that $\lim f^{[\nu(n)]}(p) = \varphi(b)$, $p \in F$, and from the identities $f \circ f^{[\nu(n)]} = f^{[\nu(n)]} \circ f$ that $\varphi(b)$ is a fixed point of f. Let g be the lifting of $f \circ \varphi$ with respect to φ satisfying $g(b) = b$. It follows that $g^{[n]}$ is the lifting of $f^{[n]} \circ \varphi$ with respect to φ satisfying the normalization $g^{[n]}(b) = b$. From $g_n = \sigma_n \circ g^{[\nu(n)]}$, where σ_n is a conformal

automorphism of Δ leaving φ invariant, we conclude that $\sigma_n(b) \to b$. Hence σ_n is the identity map for n large. It follows that $|g'(b)| < 1$. We conclude from elementary results concerning attractive fixed points that $(f^{[n]})_0^\infty$ tends pointwise to $\varphi(b)$. In fact, if V is a neighborhood of $\varphi(b)$ and K is a given compact subset of F, then $f^{[n]}(K) \subset V$ for sufficiently large n.

(2) From the equalities $g_{n+1} = h_n \circ g_n$ we conclude that $(h_n)_0^\infty$ converges to the identity map on Δ. From the definition of h_n it follows that $(f^{[\nu(n+1)-\nu(n)]})_0^\infty$ tends to the identity map on F. Since the fundamental group of F is nonabelian, it follows by the generalized "Starrheitssatz" [4, p.315] that for n large $f^{[\nu(n+1)-\nu(n)]}$ is the identity map on F. Consequently, f is a conformal automorphism of F of finite order.

There remains to show that under assumption II conclusion (iii) of Theorem 2 holds. To that end it suffices to show that if $0 < r_1, r_2 < 1$ and r_1 is so chosen that $f \circ \varphi(\{|z| < r_1\})$ and $\varphi(\{|z| < r_1\})$ are not disjoint, the connected set $U_{n \geq m} f^{[n]} \circ \varphi(\{|z| < r_1\})$ lies in $F - \varphi(\{|z| \leq r_2\})$ for m large. Otherwise there would exist an increasing sequence of positive integers $(\nu(n))_0^\infty$ and a sequence of points $(z_n)_0^\infty$, $|z_n| < r_1$, such that $f^{[\nu(n)]} \circ \varphi(z_n) \in \varphi(\{|z| \leq r_2\})$, $n = 0, 1, \ldots$. On considering liftings g_n of $f^{[\nu(n)]} \circ \varphi$ with respect to φ satisfying $|g_n(z_n)| \leq r_2$, we conclude that $f^{[\nu(n)]} \circ \varphi(0) \in \varphi(\{|z| \leq \rho\})$, $n = 0, 1, \ldots$, for some ρ satisfying $0 < \rho < 1$, contrary to assumption.

Theorem 2 follows. $\square$

We also want the corresponding result for F a nondegenerate annulus $A = \{r < |z| < r^{-1}\}$, $0 < r < 1$. The only matter that need be considered is the counterpart of (ii). The more precise known facts [3, p.469] concerning the limiting behavior of $(f^{[n]})_0^\infty$ will be subsumed by the stronger results of Theorem 1.

In the counterpart of (ii) from the fact that the sequence $(h_n)_0^\infty$ tends to the identity map on Δ we conclude that f is a conformal automorphism of A (but not necessarily of finite order). It suffices to introduce a generator τ of the group of conformal automorphisms of Δ leaving φ invariant and to note that $h_n \circ \tau = \tau \circ h_n$ for n large to conclude that $f^{[\nu(n+1)-\nu(n)]}$ is the identity map or a rotation of A onto itself for n large, whence it follows that f is a conformal automorphism of A, cf. [4].

§3 Proof of Theorem 1

It suffices to confine our attention to (iii) under the assumption that neither (i) nor (ii) holds and thereupon to the subsequent statements of the theorem.

(iii) We put down, as we may, $p_0 \in S$ such that $f^{[n]}(p_0) \neq a$, $n = 0, 1, \ldots$. By Theorem 2 we see that $p_n = f^{[n]}(p_0)$ tends to a component Γ_j of Γ. Let G_p denote the Green function for S with pole p. We note that $\lim_{n \to \infty} G_{p_n}(a) = 0$. For n a nonnegative integer, let $\mu(n)$ denote the largest nonnegative integer

m satisfying $G_{p_m}(a) \geq 2^{-n} G_{p_0}(a)$. It follows that $G_{p_{\mu(n)}}(a) > G_{p_{\mu(n)+1}}(a)$, $n = 0, 1, \ldots$. We introduce an increasing sequence $(\nu(n))_0^\infty$ of nonnegative integers such that $p_{\mu[\nu(n)]}$ tends to a point $q \in \Gamma_j$. It is to be noted that $\lim_{n \to \infty} \mu(n) = +\infty$. By the Lindelöf principle we have

$$G_{p_{\mu[\nu(n)]+1}} \circ f \geq G_{p_{\mu[\nu(n)]}}. \tag{3.1}$$

By the choice of μ it follows that

$$\frac{G_{p_{\mu[\nu(n)]+1}} \circ f}{G_{p_{\mu[\nu(n)]+1}}(a)} \geq \frac{G_{p_{\mu[\nu(n)]}}}{G_{p_{\mu[\nu(n)]}}(a)}. \tag{3.2}$$

By the principle of hyperbolic distance applied to $f^{[\mu[\nu(n)]]}$ and the points p_0 and p_1, the hyperbolic distance between $p_{\mu[\nu(n)]}$ and $p_{\mu[\nu(n)]+1}$ is bounded independently of n. On noting the behavior of φ near a point η of the unit circumference such that φ admits a continuous extension to $\Delta \cup \{\eta\}$ mapping η to q, we conclude that $p_{\mu[\nu(n)]+1} \to q$. Applying this observation to (3.2) we obtain in any case the inequality

$$u_q \circ f \geq u_q. \tag{3.3}$$

It follows that $f^{[n]}(p) \to q$ as $n \to \infty$, $p \in S$. Further, given K compact $\subset S$, the set $f^{[n]}(K)$ lies in an assigned neighborhood of q for n large as we see with the aid of the principle of hyperbolic distance.

We now show that if f is a holomorphic map of S into itself then f satisfies (1.3) if and only if f is the identity map on S. This fact shows that for the allowed f for which (3.3) was concluded the strong inequality holds. Since it is trivial that (1.3) holds for f the identity map, it suffices to proceed in the opposite direction.

If f satisfied in (1.3) and (i) of Theorem 1, then from

$$u_q \circ f^{[n]} = u_q, \quad n = 0, 1, \ldots, \tag{3.4}$$

we would conclude the constancy of u_q. This is not possible. If f were not a conformal automorphism of S, then by (3.4) and the fact that for a given compact subset K of S the set $f^{[n]}(K)$ lies in a given neighborhood of q for n large it would follow that u_q would be the real part of an analytic function w. On examining the boundary behavior of w we would conclude that w mapped S univalently onto $\{\operatorname{Re} z > 0\}$. This is not possible by the topological hypothesis on S. It follows that f is a conformal automorphism of S. Hence f admits a (unique) continuous extension to $\bar{S}$, say f^*. From the fact that $f^*(q) = q$ and the nature of f when S is an annulus ($f(z) \equiv \eta z$ or $f(z) \equiv \eta z^{-1}$, $|\eta| = 1$), resp. is the indicated region in T (f is the restriction of a conformal automorphism τ of T) we conclude that f is the identity map on S. It is to be noted that q is a fixed point of τ and that with ψ a conformal universal covering of T having domain Δ and satisfying $\psi(0) = q$, the lifting of $\tau \circ \psi$ with respect to ψ which has zero as a fixed point is the identity map on Δ.

There remains to show that if f is a holomorphic map of S into itself satisfying (1.2), then $(f^{[n]})_0^\infty$ tends to q. To that end it suffices to exclude (i) and (ii) and to note that the limit entering in (iii) is q by virtue of (1.2). Case (i) is excluded since (1.2) implies that f does not have a fixed point in S. If f were a conformal automorphism of S, the inequality (1.2) would force f^* to have q as a fixed point and it would follow as in the preceding paragraph that f would be the identity map on S. This is not possible by virtue of (1.2). $\square$

We remark that the ingenious elementary argument used by Denjoy [2] in his treatment of the iteration problem for Δ in the case where the considered function does not have a fixed point in Δ may be used to treat the result just established.

References

[1] G. Aumann and C. Carathéodory, Ein Satz über die konforme Abbildung mehrfach zusammenhängender ebene Gebiete, Math. Ann., Bd. 109, 1934, 756–763.

[2] A. Denjoy, Sur l'itération des fonctions analytiques, C.R. Acad. Sci. Paris, T. 182, 1926, 255–257.

[3] M. Heins, On the iteration of functions which are analytic and single-valued in a given multiply-connected region, Amer. J. Math. Vol. 58, no. 2, 1941, 461–480.

[4] —, A generalization of the Aumann-Carathéodory "Starrheitssatz", Duke Math. J., Vol. 8, no. 2, 1941, 312–316.

[5] —, On a problem of Heinz Hopf, J.d. Math. P.A., T. 37, 1958, 153–160.

[6] G. Valiron, Sur l'itération des fonctions holomorphes dans un demi-plan, Bull. Sc. Math., T. 55, 1931, 105–128.

[7] —, *Fonctions Analytiques*, Pr. Univ. France, Paris, 1954.

[8] J. Wolff, Sur l'itération des fonctions bornées, C.R. Acad. Sci. Paris, T. 182, 1926, 200–201.

[9] —, Sur une généralisation d'un théorème de Schwarz, C.R. Acad. Sci. Paris, T. 182, 1926, 918–920.

University of Maryland
College Park, Maryland 20742
U.S.A.

Walter Hengartner, Glenn Schober

Curvature Estimates
for some Minimal Surfaces

§1 Introduction

Let D be a simply-connected domain in the complex plane $\mathbb{C}$, and let $f = u + iv$ be a univalent, orientation-preserving, harmonic mapping from D into $\mathbb{C}$. Then f can be written in the form $f = h + \bar{g}$ where h and g belong to the linear space $H(D)$ of analytic functions on D. In addition, f can be viewed as a solution of the elliptic partial differential equation

$$\overline{f_{\bar{z}}} = a f_z \tag{1.1}$$

where the function $a = g'/h'$ belongs to $H(D)$ and satisfies $|a(z)| < 1$ for all $z \in D$. Hence the mapping f is locally quasiconformal. Conversely, any univalent solution of (1.1) with a analytic and $|a| < 1$ is an orientation-preserving harmonic mapping of D (see [4]). Observe that if ϕ is a conformal mapping from D_1 onto D, then $f_1 = f \circ \phi$ is a harmonic mapping defined on D_1 with $a_1 = a \circ \phi$.

Now, let Ω be a simply-connected domain in $\mathbb{C}$, and let S be a nonparametric surface over Ω given by $S = \{(u, v, F(u, v)) : u + iv \in \Omega\}$. Then S is a minimal surface if and only if S admits a reparametrization of the form

$$S = \{(u(z), v(z), G(z)) : z = x + iy \in D\}$$

such that $f = u + iv$ is a univalent harmonic mapping from D onto Ω and G is a real-valued harmonic function satisfying

$$G_z^2 = -a f_z^2, \tag{1.2}$$

where a is defined in (1.1). This is a conformal parametrization. More precisely, the first fundamental form for the euclidean length on S is

$$ds^2 = \rho^2 |dz|^2 \quad \text{where} \quad \rho = |h'| + |g'|.$$

Observe also that we may assume f is orientation-preserving and that we may obtain any other set of isothermal parameters by applying a conformal mapping to D. For elementary facts concerning minimal surfaces, we refer the reader to [6, 7].

From (1.2) we see that not all functions a correspond to nonparametric minimal surfaces. That is, a must be a perfect square. Thus we restrict ourselves to functions

$$a = b^2$$

with $b \in H(D)$ and $|b| < 1$ on D.

The meromorphic function i/b has geometric significance. It is the stereographic projection of the Gauss map. That is, the trace of the unit normal vector to the surface

$$\vec{n} = (2.\text{Im}\{b\},\ -2.\text{Re}\{b\},\ 1 - |b|^2)/(1 + |b|^2)$$

has stereographic projection i/b. In particular, S has a horizontal tangent plane at points where b vanishes.

If Ω is $\mathbb{C}$, then D is also $\mathbb{C}$ (see [2, Thm. 2.1]) and so b is constant by Liouville's theorem. In this case $\vec{n}$ is constant. That is, each minimal surface over $\mathbb{C}$ has to be a plane in $\mathbb{R}^3$ (Bernstein's theorem). Therefore, we assume that Ω is not $\mathbb{C}$.

Fix a point $p = p_1 + ip_2$ in Ω, and let $P = (p_1, p_2, F(p))$ be the corresponding point of S. We may choose D to be the unit disk $\mathbb{D} = \{z\ :\ |z| < 1\}$ and normalize in such a way that the harmonic mapping f satisfies $f(\mathbb{D}) = \Omega$ and $f(0) = p$. Then the Gaussian curvature $k(P)$ of S at the point P is

$$\left. \begin{aligned} k(P) &= \frac{-(\Delta \log \rho)(0)}{\rho(0)^2} = \frac{-|h'(0)g''(0) - h''(0)g'(0)|^2}{|h'(0)g'(0)|(|h'(0)| + |g'(0)|)^4} \\ &= \frac{-4|b'(0)|^2}{(1 + |b(0)|^2)^4 |h'(0)|^2}. \end{aligned} \right\} \tag{1.3}$$

Many estimates of $|k(P)|$ for nonparametric minimal surfaces have been established. If $\Omega = \{w\ :\ |w| < r\}$ and $p = 0$, then the estimate $|k(P)| \leq 16\pi^2/(27r^2)$ follows from R.R. Hall's work [3]. Unfortunately, it is not sharp. However, if one assumes that S has a horizontal tangent plane at P, then the sharp estimate $|k(P)| \leq \pi^2/(2r^2)$ is known. For a general domain Ω one has under the assumption of a horizontal tangent plane at P the sharp estimate $|k(P)| \leq 64/(9d^2)$ where d is the distance along the surface from P to ∂S. These and a number of other estimates can be found in [1, 7].

In this note we shall give sharp estimates of $|k(P)|$ for the following three cases:

(1) $\qquad\qquad \Omega_1 = \{w \,:\, |\mathrm{Im}\{w\}| < \pi/4\}, \quad$ arbitrary $\quad p \in \Omega_1;$

(2) $\qquad\qquad \Omega_2 = \{w \,:\, \mathrm{Im}\{w\} > 0\}, \qquad$ arbitrary $\quad p \in \Omega_2;$

(3) $\qquad\qquad \Omega_3 = \mathbb{C} \setminus (-\infty, 0], \qquad\quad$ arbitrary $\quad p > 0.$

§2 General framework

Assume that the domain Ω is convex in the u-direction, that is, the intersection of Ω with each horizontal line is connected. This is the case for the three domains listed above. Let $f = h + \bar{g}$, $g, h \in H(\mathbf{D})$, be the associated univalent harmonic mapping of $\mathbf{D}$ onto Ω. By a result of J.G. Clunie and T. Sheil-Small [2, p.14], the analytic function $\varphi = h - g$ is then univalent on $\mathbf{D}$, and its image is also a domain convex in the u-direction. We are free to normalize $g(0) = 0$, in which case $\varphi(0) = h(0) = f(0) = p$.

From $\mathrm{Im}\{f\} = \mathrm{Im}\{\varphi\}$, $\mathrm{Re}\{f\} = \mathrm{Re}\{h + g\}$, and $h' + g' = \frac{1+b^2}{1-b^2}\varphi'$ it follows that we have the representation

$$f(z) = \mathrm{Re}\left\{ p + \int_0^z \frac{1+b^2}{1-b^2}\varphi' dz \right\} + i\,\mathrm{Im}\{\varphi\}. \qquad (2.1)$$

In addition, the equation (1.2) leads to

$$G(z) = \pm\mathrm{Im}\left\{ \int_0^z \frac{2b}{1-b^2}\varphi' dz \right\} + \text{constant}. \qquad (2.2)$$

Since $h' = \varphi'/(1 - b^2)$, the Gaussian curvature (1.3) becomes

$$k(P) = \frac{-4|1 - b(0)^2|^2 |b'(0)|^2}{(1 + |b(0)|^2)^4 |\varphi'(0)|^2}.$$

Furthermore, the elementary estimate $|b'(0)| \leq 1 - |b(0)|^2$ from Schwarz's lemma and the triangle inequality imply

$$|k(P)| \leq \frac{4|1 - b(0)^2|^2(1 - |b(0)|^2)^2}{(1 + |b(0)|^2)^4 |\varphi'(0)|^2} \leq \frac{4}{|\varphi'(0)|^2}. \qquad (2.3)$$

The first estimate is sharp only if b is a Möbius transformation, and the second one is sharp if and only if $b(z) = sz$ with $|s| = 1$.

It will be convenient to define the function

$$\Xi_s(z) = \int\limits_0^z \frac{\varphi'(z)}{1 - sz} \, dz. \tag{2.4}$$

Then for $b(z) = sz$, $|s| = 1$, the mapping (2.1) and equation (2.2) lead to the parametrization

$$\left. \begin{aligned} u(z) &= \mathrm{Re}\{\Xi_s(z) + \Xi_{-s}(z) - \Xi_0(z)\} + p_1 \\ v(z) &= \mathrm{Im}\{\Xi_0(z)\} + p_2 \\ G(z) &= \mathrm{Im}\{\Xi_s(z) - \Xi_{-s}(z)\} + \text{constant} \end{aligned} \right\} \tag{2.5}$$

for the minimal surface.

In order to derive useful estimates from (2.3), we shall introduce properties of the functions φ that depend on the domains Ω.

§3   The case of a strip-domain $\varOmega_1$

Our first example deals with nonparametric minimal surfaces S which lie over the horizontal strip $\Omega_1 = \{w \,:\, |\mathrm{Im}\{w\}| < \pi/4\}$ and with an arbitrary point $p = p_1 + ip_2$ of Ω_1.

Since $f(\boldsymbol{D}) = \Omega_1$ and $\mathrm{Im}\{\varphi\} = \mathrm{Im}\{f\}$, the function φ maps $\boldsymbol{D}$ conformally into Ω_1. Furthermore, with possibly two exceptions, the boundary values of φ are on $\partial\Omega_1$. Thus $\varphi(z) = \varphi_1(e^{i\alpha}z)$ where

$$\varphi_1(z) = p + \frac{1}{2} \log\left[\frac{1 - \zeta z}{1 - z}\right], \quad \zeta = -e^{-4ip_2}, \tag{3.1}$$

maps $\boldsymbol{D}$ onto Ω_1 with $\varphi_1(0) = p$. Since a rotation of the disk $\boldsymbol{D}$ simply reparametrizes the same surface, it is no loss of generality to assume that $e^{i\alpha} = 1$. Consequently, the representation (2.1) becomes

$$f(z) = \frac{1}{2}\mathrm{Re}\left\{\int\limits_0^z \frac{[1 + b(z)^2](1 - \zeta)dz}{[1 - b(z)^2](1 - \zeta z)(1 - z)}\right\} + \frac{i}{2}\arg\left[\frac{1 - \zeta z}{1 - z}\right] + p, \quad (3.2)$$

where $\zeta = -e^{-4ip_2}$, $b \in H(\boldsymbol{D})$, and $|b| < 1$. Note that $|\zeta| = 1$, but $\zeta \neq 1$.

Conversely, each function (3.2) is a univalent harmonic mapping of $\boldsymbol{D}$ into Ω_1 whose boundary cluster sets are singletons which lie on $\partial\Omega_1$, except possibly at 1 and $\bar\zeta$ (for a finer description see [4, 5]).

Since $|\varphi'(0)| = \cos(2p_2)$, the estimate (2.3) for the Gaussian curvature becomes $|k(P)| \le 4\sec^2(2p_2)$. In other words, we have the following.

Theorem 3.1. *If S is a nonparameteric minimal surface over the strip-domain $\Omega_1 = \{w \; : \; |\mathrm{Im}\{w\}| < \pi/4\}$ and if $p = p_1 + ip_2$ is an arbitrary point of Ω_1, then we have the sharp estimates*

$$|k(P)| \le 4 \quad \text{if} \quad p_2 = 0 \quad \text{and} \quad |k(P)| < 4\sec^2(2p_2) \quad \text{if} \quad p_2 \ne 0. \qquad (3.3)$$

It remains to study the case of equality in (3.3) and to show that the inequalities are best possible for each p in Ω_1. Indeed, equality occurs in the estimates (2.3) and (3.3) when $b(z) = sz$ with $|s| = 1$. Thus we need to study the surfaces generated by (2.5).

Using (3.1) the function (2.4) becomes

$$2\Xi_s(z) = \frac{s(1-\zeta)}{(\zeta - s)(s-1)} \log(1 - sz) + \frac{\zeta}{\zeta - s} \log(1 - \zeta z) + \frac{1}{s-1} \log(1 - z)$$

for $s \ne 1, \zeta$ and

$$2\Xi_1(z) = \frac{z}{1-z} + \frac{\zeta}{\zeta - 1} \log\left[\frac{1-\zeta z}{1-z}\right], \quad 2\Xi_\zeta(z) = \frac{1}{1-\zeta} \log\left[\frac{1-\zeta z}{1-z}\right] - \frac{\zeta z}{1-\zeta z}.$$

Thus the parametrization (2.5) becomes

$$u(z) = \frac{1}{2}\mathrm{Re}\left\{ \frac{s(1-\zeta)}{(\zeta - s)(s-1)} \log(1 - sz) + \frac{s(1-\zeta)}{(\zeta + s)(s+1)} \log(1 + sz) \right.$$

$$\left. + \frac{\zeta^2 + s^2}{\zeta^2 - s^2} \log(1 - \zeta z) + \frac{s^2 + 1}{s^2 - 1} \log(1 - z) \right\} + p_1,$$

$$v(z) = \frac{1}{2}\arg\left[\frac{1-\zeta z}{1-z}\right] + p_2, \qquad (3.4)$$

$$G(z) = \frac{1}{2}\mathrm{Im}\left\{ \frac{s(1-\zeta)}{(\zeta - s)(s-1)} \log(1 - sz) - \frac{s(1-\zeta)}{(\zeta + s)(s+1)} \log(1 + sz) \right.$$

$$\left. + \frac{2\zeta s}{\zeta^2 - s^2} \log(1 - \zeta z) + \frac{2s}{s^2 - 1} \log(1 - z) \right\} + \text{constant},$$

at least when $s \ne \pm 1, \pm\zeta$.

Let us first consider the case when p is real. Then $p_2 = 0$, $\zeta = -1$, and (3.4) becomes

$$\left.\begin{aligned}
u(z) &= \frac{1}{2\mathrm{Im}\{s\}} \arg\left[\frac{1+sz}{1-sz}\right] - \frac{\mathrm{Re}\{s\}}{2\mathrm{Im}\{s\}} \arg\left[\frac{1+z}{1-z}\right] + p_1 \\[2mm]
v(z) &= \frac{1}{2} \arg\left[\frac{1+z}{1-z}\right] \\[2mm]
G(z) &= \frac{1}{2\mathrm{Im}\{s\}} \log\left|\frac{1-s^2 z^2}{1-z^2}\right| + \text{constant}
\end{aligned}\right\} \tag{3.5}$$

when $|s| = 1$, $s \neq \pm 1$, and

$$u(z) = \mathrm{Re}\left\{\frac{z}{1-z^2}\right\} + p_1, \quad v(z) = \frac{1}{2}\arg\left[\frac{1+z}{1-z}\right],$$

$$G(z) = \pm\mathrm{Im}\left\{\frac{z^2}{1-z^2}\right\} + \text{constant}$$

if $s = \pm 1$. In the latter case $s = \pm 1$, we obtain a minimal surface over all of Ω_1. Indeed, put $\frac{1+z}{1-z} = Re^{it}$. Then $R > 0$, $-\pi/2 < t < \pi/2$, and we have: $u = \frac{1}{4}(R-\frac{1}{R})\cos(t)+p_1$, $v = t/2$, and $G = \pm\frac{1}{4}(R-\frac{1}{R})\sin(t)+\text{constant}$. In this case, u varies from $-\infty$ to $+\infty$ on each horizontal line $v = \text{constant}$. These minimal surfaces are *helicoids* and can be expressed by the nonparametric equations $G = F(u,v) = \pm(u - p_1)\tan(2v) + \text{constant}$. Thus the estimate (3.3) is sharp and attained by helicoids when p is real.

Let us consider further the case where p is real, but $s \neq \pm 1$. Then (3.5) represents minimal surfaces S with $k(P) = -4$ which do not lie over all of Ω_1, but rather over quadrilaterals whose vertices are on $\partial\Omega_1$. However, they are limits of surfaces over all of Ω_1 in the following sense. If in (3.4) we replace s by $(1 - \frac{1}{n})s$, then we obtain a sequence of minimal surfaces S_n over all of Ω_1 which converges locally uniformly in the parameter disk $\mathbb{D}$ to S and whose Gaussian curvatures $k_n(P)$ converge to -4 as $n \to \infty$. As a special case let us consider $s = i$. Then (3.5) becomes

$$u(z) = \frac{1}{2}\arg\left[\frac{1+iz}{1-iz}\right] + p_1, \quad v(z) = \frac{1}{2}\arg\left[\frac{1+z}{1-z}\right],$$

$$G(z) = \frac{1}{2}\log\left|\frac{1+z^2}{1-z^2}\right| + \text{constant}$$

and represents a minimal surface that lies over the square with the four vertices $p_1 \pm \pi(1 \pm i)/4$. It is known as *Scherk's surface* and can be expressed by the nonparametric equation $G = F(u,v) = \frac{1}{2}\log\left[\frac{\cos(2v)}{\cos(2(u-p_1))}\right]$.

Now let $p = p_1 + ip_2 \in \Omega_1$ with $p_2 \neq 0$ and $\zeta = -e^{-4ip_2}$ as before. Then equations (3.4) define minimal surfaces S with Gaussian curvature $k(P) = -4\sec^2(2p_2)$, but

$$u(z) = \frac{-1}{2}\left[\operatorname{Im}\left\{\frac{s(1-\zeta)}{(\zeta-s)(s-1)}\right\}\arg(1-sz)\right.$$

$$+ \operatorname{Im}\left\{\frac{s(1-\zeta)}{(\zeta+s)(s+1)}\right\}\arg(1+sz)$$

$$\left.+ \frac{\operatorname{Re}\{\bar{\zeta}s\}}{\operatorname{Im}\{\bar{\zeta}s\}}\arg(1-\zeta z) - \frac{\operatorname{Re}\{s\}}{\operatorname{Im}\{s\}}\arg(1-z)\right] + p_1$$

is bounded for $s \neq \pm 1, \pm\zeta$. Since $\zeta \neq \pm 1$, one can show that $u(z)$ is bounded either above or below in all other cases, too. That is, none of these surfaces lies over the whole strip Ω_1. Therefore equality does not occur in our curvature estimate. However, if in (3.4) we replace s by $(1 - \frac{1}{n})s$, then we get again a sequence of minimal surfaces S_n over all of Ω_1 which converges locally uniformly in the parameter disk $\boldsymbol{D}$ to S and whose Gaussian curvatures $k_n(P)$ converge to $-4\sec^2(2p_2)$ as $n \to \infty$. Therefore, (3.3) is best possible.

§4   The case of a half-plane Ω_2

Let Ω now be the upper half-plane $\Omega_2 = \{w \; : \; \operatorname{Im}\{w\} > 0\}$, and let $p = p_1 + ip_2 \in \Omega_2$. The procedure to obtain a sharp estimate for the Gaussian curvature $k(P)$ is the same as for the strip above. We just replace φ_1 by the mapping

$$\varphi_2(z) = \frac{p - \bar{p}z}{1 - z} \tag{4.1}$$

of $\boldsymbol{D}$ onto Ω_2 with $\varphi_2(0) = p$. Since $|\varphi'(0)| = 2p_2$, the curvature estimate (2.3) becomes $|k(P)| \leq 1/(p_2)^2$. In other words we have the following.

Theorem 4.1 *If S is a nonparametric minimal surface over the half-plane $\Omega_2 = \{w \; : \; \operatorname{Im}\{w\} > 0\}$ and if $p = p_1 + ip_2 \in \Omega_2$, then we have the sharp estimate*

$$|k(P)| \leq 1/(p_2)^2. \tag{4.2}$$

It remains to show that this estimate is best possible for each $p \in \Omega_2$. As before, the estimate is sharp only for the surfaces parametrized by (2.5) with $|s| = 1$. With (4.1) the function (2.4) becomes

$$\Xi_s(z) = 2ip_2\left[\frac{z}{(1-s)(1-z)} + \frac{s}{(1-s)^2}\log\left[\frac{1-z}{1-sz}\right]\right]$$

for $s \neq 1$ and $\Xi_1(z) = \frac{ip_2 z(2-z)}{(1-z)^2}$. Therefore the parametrization (2.5) is

$$u(z) = p_1 - p_2 \left[\frac{2\mathrm{Re}\{s\}}{\mathrm{Im}\{s\}} \mathrm{Re}\left\{ \frac{z}{1-z} \right\} + \frac{1}{\mathrm{Re}\{s\} - 1} \arg\left[\frac{1-z}{1-sz} \right] \right.$$
$$\left. - \frac{1}{\mathrm{Re}\{s\} + 1} \arg\left[\frac{1-z}{1+sz} \right] \right]$$

$$v(z) = p_2 \mathrm{Re}\left\{ \frac{1+z}{1-z} \right\}$$

$$G(z) = p_2 \left[\frac{-2}{\mathrm{Im}\{s\}} \mathrm{Im}\left\{ \frac{z}{1-z} \right\} + \frac{1}{\mathrm{Re}\{s\} - 1} \log\left| \frac{1-z}{1-sz} \right| \right.$$
$$\left. + \frac{1}{\mathrm{Re}\{s\} + 1} \log\left| \frac{1-z}{1+sz} \right| \right] + \text{constant}$$

if $|s| = 1$ and $s \neq \pm 1$, and

$$u(z) = p_1 - p_2 \left[\mathrm{Im}\left\{ \frac{z}{(1-z)^2} \right\} + \frac{1}{2} \arg\left[\frac{1+z}{1-z} \right] \right]$$

$$v(z) = p_2 \mathrm{Re}\left\{ \frac{1+z}{1-z} \right\}$$

$$G(z) = \pm p_2 \left[\mathrm{Re}\left\{ \frac{z}{(1-z)^2} \right\} - \frac{1}{2} \log\left| \frac{1+z}{1-z} \right| \right] + \text{constant}$$

if $s = \pm 1$.

Consider first the case $s = \pm 1$. Putting $\frac{1 \pm z}{1-z} = Re^{it}$, we have $R > 0$, $-\pi/2 < t < \pi/2$, and

$$u = p_1 - \frac{1}{4} p_2 [2t + R^2 \sin(2t)], \quad v = p_2 R \cos(t),$$

$$G = \pm \frac{1}{4} p_2 [R^2 \cos(2t) - \log(R^2)] + \text{constant}.$$

It is evident that u varies from $-\infty$ to $+\infty$ on each horizontal line, and therefore, the two minimal surfaces lie over all of Ω_2. Thus the estimate (4.2) is attained and sharp.

If $s \neq \pm 1$, then the corresponding minimal surface S does not lie over all of Ω_2. But we may, as in the cases before, approach S uniformly on compact subsets of the parameter disk by minimal surfaces over Ω_2. In the particular case of $s = i$, we get

$$u = p_1 - p_2 \arg(1 + R^2 e^{2it}), \quad v = p_2 R \cos(t),$$

$$G = \frac{1}{2}p_2 \left[\log \left[\frac{R^2 + 2R\sin(t) + 1}{R^2 - 2R\sin(t) + 1} \right] - R\sin(t) \right] + \text{constant}.$$

Since

$$\tan \left[\frac{u - p_1}{p_2} \right] = \frac{-R^2 \sin^2(2t)}{1 + R^2 \cos^2(2t)} = \frac{-2p_2 v(R\sin(t))}{(p_2)^2 + v^2 - (p_2)^2(R\sin(t))^2},$$

one can solve explicitly for $R\sin(t)$ in terms of u and v, substitute into the expression for G, and obtain an explicit nonparametric representation for this surface.

Remark. For the strip Ω_1 and half-plane Ω_2 the extremal surfaces were all generated by harmonic mappings with b of the form $b(z) = sz$, $|s| = 1$. Since $b(0) = 0$, the normal vector to the surface at P is vertical. In addition, since $|b(z)| \to 1$ as $|z| \to 1$, the normal vector becomes horizontal as one approaches the boundary of the surface, even in $\overline{I\!R^3} = I\!R^3 \cup \{\infty\}$. In other words, we have the somewhat surprising observation that the minimal surfaces for which $|k(P)|$ is largest have tangent planes that are horizontal at P and more and more vertical as one approaches the boundary. More generally, whenever we have a sequence of minimal surfaces S_n over Ω_1 or Ω_2 for which the Gaussian curvature at a given point P converges to its minimum value, then there is a subsequence which converges locally uniformly in the parameter disk $I\!\!D$ to a minimal surface S with these same properties. However, S may lie only over a subdomain of Ω_1 or Ω_2.

§5 The case of a slit plane Ω_3

Let Ω be the slit plane $\Omega_3 = \mathbb{C} \setminus (-\infty, 0]$, and restrict p to be a point of the positive real axis.

Since $\text{Im}\{\varphi\} = \text{Im}\{f\}$, the function $\varphi = h - g$ maps $I\!\!D$ onto the plane with slits on the real axis. That is, φ is of the form $\varphi(z) = \varphi_3(e^{i\alpha}z)$ where

$$\varphi_3(z) = p + \frac{\lambda z}{1 - 2\xi z + z^2}$$

and $\lambda > 0$, $-1 \leq \xi \leq 1$. A rotation of the disk $I\!\!D$ does not affect the surface, and so we may assume that $e^{i\alpha} = 1$ and $\varphi = \varphi_3$. Thus the real part of the mapping (2.1) has the representation

$$u(z) = p + \lambda \text{Re} \left\{ \int_0^z \frac{[1 + b(z)^2](1 - z^2)dz}{[1 - b(z)^2](1 - 2\xi z + z^2)^2} \right\}, \tag{5.0}$$

and using $|\varphi'(0)| = \lambda$, we will find it convenient to write the curvature estimate (2.3) in the form

$$|k(P)| \leq \frac{4|1 - b(0)^2|^2(1 - |b(0)|^2)^2}{\lambda^2(1 + |b(0)|^2)^4}.$$

(5.1)

Equality occurs only if b is a Möbius transformation.

On the interval $0 < x < 1$ the function $u(x)$ must increase from p to $+\infty$. This can occur only if $\xi = 1$, for otherwise the integrand and integral (5.0) are bounded on this interval.

On the interval $-1 < x < 0$ the function $u(x)$ increases from 0 to p; in particular,

$$0 = p + \lambda \mathrm{Re}\left\{ \int_0^{-1} \frac{[1 + b(x)^2](1 - x^2)dx}{[1 - b(x)^2](1 - 2\xi x + x^2)^2} \right\}$$

where $\xi = 1$. Solving this for λ, we may rewrite the curvature estimate (5.1) as

$$|k(P)| \leq \frac{4|1 - b(0)^2|^2(1 - |b(0)|^2)^2}{p^2(1 + |b(0)|^2)^4}.$$
$$\cdot \left[\int_{-1}^0 \mathrm{Re}\left\{ \frac{1 + b(x)^2}{1 - b(x)^2} \right\} \frac{(1 + x)}{(1 - x)^3} dx \right]^2.$$

(5.2)

In order to estimate the integrand, the following lemmas will be useful.

Lemma 5.1. *Let $b \in H(\boldsymbol{D})$ satisfy $|b| < 1$ on $\boldsymbol{D}$. Then we have*

$$\mathrm{Re}\left\{ \frac{1 + b(z)^2}{1 - b(z)^2} \right\} \leq \frac{1}{2}\left[M\left[\frac{1 + |z|}{1 - |z|} \right] + \frac{1}{M}\left[\frac{1 - |z|}{1 + |z|} \right] \right]$$

(5.3)

for all $z \in \boldsymbol{D}$ where $M = \max\left\{ \frac{1 - |b(0)|^2}{|1 - b(0)|^2}, \frac{|1 - b(0)|^2}{1 - |b(0)|^2} \right\}$. This inequality is sharp for all real values $z \in \boldsymbol{D}$ if and only if $b(z) = \pm\frac{z + \sigma}{1 + \sigma z}$, $-1 < \sigma < 1$.

Proof. Put $W = U + iV = \frac{1 + b}{1 - b}$. Then $\mathrm{Re}\left\{ \frac{1 + b^2}{1 - b^2} \right\} = \frac{1}{2}\mathrm{Re}\{W + \frac{1}{W}\} \leq \frac{1}{2}(U + \frac{1}{U})$. Using the fact that $\Psi(U) = U + \frac{1}{U}$ is a convex function of U on $\boldsymbol{R}^+$ and the inequality

$$U(0)\frac{1 - |z|}{1 + |z|} \leq U(z) \leq U(0)\frac{1 + |z|}{1 - |z|}$$

(5.4)

for positive harmonic functions, we conclude that

$$\mathrm{Re}\left\{\frac{1+b(z)^2}{1-b(z)^2}\right\} \le \frac{1}{2}\mathrm{Max}\left\{U(0)\frac{1-|z|}{1+|z|}+\frac{1}{U(0)}\frac{1+|z|}{1-|z|},\right.$$

$$\left. U(0)\frac{1+|z|}{1-|z|}+\frac{1}{U(0)}\frac{1-|z|}{1+|z|}\right\}$$

$$= \frac{1}{2}\left[M\left[\frac{1+|z|}{1-|z|}\right]+\frac{1}{M}\left[\frac{1-|z|}{1+|z|}\right]\right].$$

Equality in (5.4) for some real $z \in \boldsymbol{D} \setminus \{0\}$ occurs only for functions of the form $U(z) = \gamma \mathrm{Re}\left\{\frac{1\pm z}{1\mp z}\right\}$, $\gamma > 0$, which correspond to the indicated functions b. Hence, equality in (5.3) for all real $z \in \boldsymbol{D}$ can occur only for these functions, and one easily verifies that equality does occur.

Lemma 5.2. *For positive constants α and β define the function*

$$H(z) = \frac{|1-z^2|}{(1+|z|^2)^2}\left[\alpha\frac{(1-|z|^2)^2}{|1-z|^2}+\beta|1-z|^2\right]. \tag{5.5}$$

Then H assumes a maximum over $\boldsymbol{D}$ at the origin if $\alpha = \beta$ and at $z_o = B[1-\sqrt{1-B^{-2}}]$ if $\alpha \ne \beta$, where $B = A[1+\sqrt{1+2A^{-2}}]$ and $A = \frac{1}{2}\left[\frac{\alpha+\beta}{\alpha-\beta}\right]$. The maximum value is $H(0) = \alpha + \beta$ if $\alpha = \beta$ and $H(z_o) = \frac{(\alpha+\beta)[1-B^{-2}]^{3/2}}{1-2B^{-2}}$ if $\alpha \ne \beta$.

Proof. We first compute

$$(1-z^2)(1+|z|^2)^3\left|\frac{1-z}{1+z}\right|H_z(z) = \alpha(1-|z|^2)(1-4\bar{z}+4z|z|^2-|z|^4)$$

$$-\beta|1-z|^4(1+4\mathrm{Re}\{z\}+|z|^2).$$

It is apparent that H_z can vanish at a point of $\boldsymbol{D}$ only if the term $-4\bar{z}+4z|z|^2$ is real, and this occurs only if z is real. Therefore H can have a maximum in $\boldsymbol{D}$ only on the real axis.

Consider $H(x) = \frac{(1-x^2)}{(1+x^2)^2}[\alpha(1+x)^2+\beta(1-x)^2]$ for $-1 < x < 1$. If $\alpha = \beta$, it is elementary that $H'(x) = 0$ only for $x = 0$ and that $H(0) = \alpha + \beta$. Now assume that $\alpha \ne \beta$. Then $H'(x) = 0$ only if

$$(\alpha-\beta)x^4 - 2(\alpha+\beta)x^3 - 6(\alpha-\beta)x^2 - 2(\alpha+\beta)x + (\alpha-\beta) = 0.$$

In factored form this equation is equivalent to

$$(x^2 - 2Cx + 1)(x^2 - 2Dx + 1) = 0$$

where $C = A + \sqrt{A^2 + 2}$ and $D = A - \sqrt{A^2 + 2}$. Since $|A| > \frac{1}{2}$, the only root in the interval $-1 < x < 1$ is $z_o = B[1 - \sqrt{1 - B^{-2}}]$, and lengthy computations lead to the representations $H(z_o) = \frac{(\alpha+\beta)[1-B^{-2}]^{3/2}}{1-2B^{-2}}$ and $H(z_o) = \beta\Phi(B)$ where $\Phi(B) = \frac{2(B-1)\sqrt{1-B^{-2}}}{B-2}$.

To show that this value is indeed a maximum over $\boldsymbol{D}$, we shall compare it to the values of H at $\partial\boldsymbol{D}$. If we define $H(1) = 0$, then H is continuous on $\bar{\boldsymbol{D}}$ and $H(e^{it}) = \beta|\sin(t)|\,|1 - \cos(t)|$. The maximum of this expression is $\frac{3\sqrt{3}}{4}\beta$ and occurs when $\cos(t) = -1/2$. If $\alpha = \beta$, it is clear that $H(0) = \alpha + \beta$ is larger than $\frac{3\sqrt{3}}{4}\beta$. If $\alpha \neq \beta$, it remains to show that $H(z_o) = \beta\Phi(B)$ is larger than $\frac{3\sqrt{3}}{4}\beta$.

Since $|A| > \frac{1}{2}$, we are concerned with $|B| > 2$. For these values the derivative $\Phi'(B) = \frac{4(1-B)(2+B^2)}{B^3(B-2)^2\sqrt{1-B^{-2}}}$ is negative. Therefore $\Phi(B) > \Phi(-2) = \frac{3\sqrt{3}}{4}$ if $B < -2$ and $\Phi(B) > \Phi(+\infty) = 2 > \frac{3\sqrt{3}}{4}$ if $B > 2$. In any case, $H(z_o)$ provides the largest value.

Theorem 5.1. *If S is a nonparametric minimal surface over the slit plane $\Omega_3 = \boldsymbol{C} \setminus (-\infty, 0]$ and if $p > 0$, then we have the sharp estimate*

$$|k(P)| \leq \frac{3 + 2\sqrt{3}}{12p^2}. \tag{5.6}$$

Proof. Combining the curvature bound (5.2) with Lemma 5.1, we obtain

$$\left.\begin{aligned}
|k(P)| &\leq \frac{|1 - b(0)^2|^2(1 - |b(0)|^2)^2}{p^2(1 + |b(0)|^2)^4} \cdot \\
&\quad \cdot \left[\int_{-1}^{0} \left[M\left[\frac{1-x}{1+x}\right] + \frac{1}{M}\left[\frac{1+x}{1-x}\right]\right]\frac{(1+x)}{(1-x)^3}\,dx\right]^2 \\
&= \frac{|1 - b(0)^2|^2(1 - |b(0)|^2)^2}{p^2(1 + |b(0)|^2)^4}\left[\frac{M}{2} + \frac{1}{6M}\right]^2.
\end{aligned}\right\} \tag{5.7}$$

If we refer to the function (5.5) as $H(z; \alpha, \beta)$, then this estimate is

$$|k(P)| \leq \frac{1}{36p^2}[H(b(0); \alpha, \beta)]^2$$

where $\alpha = 3$ and $\beta = 1$ if $1 - |b(0)|^2 \geq |1 - b(0)|^2$ and where $\alpha = 1$ and $\beta = 3$ if $1 - |b(0)|^2 < |1 - b(0)|^2$. By Lemma 5.2, the maximum of both $H(\cdot; 3, 1)$ and $H(\cdot; 1, 3)$ is $\sqrt{9 + 6\sqrt{3}}$. Thus we have the estimate (5.6).

Equality in (5.1) occurs only if b is a Möbius transformation, and by Lemma 5.1, equality in (5.7) occurs only if b is of the form $b(z) = \pm\frac{z+\sigma}{1+\sigma z}$, $-1 < \sigma < 1$. Finally, using Lemma 5.2, we find that equality in the final estimate occurs if and only if $\sigma = \pm[1 + \sqrt{3} - \sqrt{3 + 2\sqrt{3}}]$.

With these choices for b, the parametrization from (2.1) and (2.2) for the extremal surfaces becomes

$$u(z) = p + \frac{\lambda}{1 - \sigma^2}\mathrm{Re}\left\{\frac{(1 + \sigma)^2 z^2(3 - z)}{3(1 - z)^3} + \frac{(1 + \sigma^2)z}{1 - z}\right\}$$

$$v(z) = \lambda\mathrm{Im}\left\{\frac{z}{(1 - z)^2}\right\}$$

$$G(z) = \frac{\pm 2\lambda}{1 - \sigma^2}\mathrm{Im}\left\{\frac{(1 + \sigma)^2 z^2(3 - z)}{6(1 - z)^3} + \frac{\sigma z}{1 - z}\right\} + \text{constant}.$$

The constant λ is determined by $u(-1) = 0$; that is, $\lambda = \frac{3(1-\sigma^2)p}{1-\sigma+\sigma^2}$. In terms of $Z = X + iY = \frac{1+z}{1-z}$, $X > 0$, the surfaces are described by

$$u(Z) = \frac{p}{4(1 - \sigma + \sigma^2)}[(1 + \sigma)^2\mathrm{Re}\{Z^3\} + 3(1 - \sigma)^2\mathrm{Re}\{Z\}]$$

$$v(Z) = \frac{3p(1 - \sigma^2)}{4(1 - \sigma + \sigma^2)}\mathrm{Im}\{Z^2\}$$

$$G(Z) = \frac{\pm p}{4(1 - \sigma + \sigma^2)}[(1 + \sigma)^2\mathrm{Im}\{Z^3\} - 3(1 - \sigma)^2\mathrm{Im}\{Z\}] + \text{constant}.$$

To see that these surfaces lie over all of Ω_3, consider the horizontal line $v = c$. This line corresponds to the curve $Y(X) = \frac{2c(1-\sigma+\sigma^2)}{3p(1-\sigma^2)X}$. Now for $X > 0$ one sees that

$$u = \frac{p(1 + \sigma)^2}{4(1 - \sigma + \sigma^2)}X^3 + \frac{3p(1 - \sigma)^2}{4(1 - \sigma + \sigma^2)}X - \frac{c^2(1 - \sigma + \sigma^2)}{3p(1 - \sigma)^2}\frac{1}{X}$$

varies from $-\infty$ to $+\infty$ if $c \neq 0$ and varies from 0 to $+\infty$ if $c = 0$.

References

[1] Y. Abu-Muhanna and G. Schober, "Harmonic mappings onto convex domains", Canadian Math. J., 39 (1987), 1489–1530.

[2] J.G. Clunie and T. Sheil-Small, "Harmonic univalent functions", Ann. Acad. Sci. Fenn. Ser. A.I. 9 (1984), 3–25.

[3] R.R. Hall, "On an inequality of E. Heinz", J. Analyse Math. 42 (1982/83), 185–198.

[4] W. Hengartner and G. Schober, "Harmonic mappings with given dilatation", J. London Math. Soc. 33 (1986), 473–483.

[5] W. Hengartner and G. Schober, "Univalent harmonic functions", Trans. Amer. Math. Soc. 299 (1987), 1–31.

[6] J.C.C. Nitsche, *Vorlesungen über Minimalflächen*, Springer-Verlag, 1975.

[7] R. Osserman, *A Survey of Minimal Surfaces*, Dover, 1986.

This work was supported in part by grants from the National Research Council (Canada) and the National Science Foundation (USA).

Université Laval   Indiana University
Quebec, P.Q., Canada   Bloomington, Indiana, USA

Joseph Hersch

On some elementary Applications of the Reflection Principle to Schwarz-Christoffel Integrals

§1 Introduction

We shall consider here only *one-to-one conformal mappings*. In the application of the Schwarz-Christoffel formula for mapping the upper half-plane G_z onto a polygon G_w, the main difficulty is to find the (real) pre-images of the vertices of G_w. We want to emphasize here how the reflection principle may open an elementary path from a mapping $G_z \longrightarrow G_w$ to a mapping $G_z \longrightarrow \hat{G}_w$, where $\hat{G}_w$ is an extension of G_w by reflection across one or several boundary segments.

By the reflection principle, the mapping $w : G_z \longrightarrow G_w$ can be extended to a mapping, *written again w*, of a domain $\hat{G}_z$ onto $\hat{G}_w$. If G_w is extended to $\hat{G}_w$ by reflection across *one* boundary segment, then $\hat{G}_z$ is in the z-plane. But if we have a reflection across *several* boundary segments, then $\hat{G}_z$ is *on a covering surface* above the z-plane.

If we know a mapping $G_z \xrightarrow{f} \hat{G}_z$, then we consider $G_z \xrightarrow{f} \hat{G}_z \xrightarrow{w} \hat{G}_w$, i.e. the function

$$\hat{w}(z) = w \circ f(z) \tag{1}$$

maps G_z onto $\hat{G}_w$. — The function $f(z)$ will be rational.

§2 Extension of G_w to $\hat{G}_w$ by reflection across one boundary segment γ_w.

2.1. *If the corresponding segment γ_z ($\subset \partial G_z$) is the real segment $-\infty < x < 0$, then $\hat{G}_z$ is the z-plane with the real slit $0 \leq x \leq \infty$. The function $z \longrightarrow \hat{z} = f(z) = z^2$ maps G_z onto $\hat{G}_z$.* Formula (1) gives the function

$\hat{w}(z) = w(z^2)$, mapping the upper half-plane G_z onto the "double polygon" $\hat{G}_w$. This trivial formula is the simplest introduction to what follows. Its application to Schwarz-Christoffel integrals is straightforward, since $\hat{w}'(z) = w'(z^2) \cdot 2z$.

2.2. *If the corresponding segment γ_z is the real segment $-1 < x < 1$, then $\hat{G}_z$ is the z-plane with the two real slits $-\infty \le x \le -1$ and $1 \le x \le \infty$.* The Joukowski function $z \longrightarrow \hat{z} = f(z) = (z + z^{-1})/2$ maps G_z onto $\hat{G}_z$. We again use (1) and apply its derivative to Schwarz-Christoffel integrals.

§3   Extension across two consecutive boundary segments

3.1. We shall now extend the upper half-plane G_z to $\hat{G}_z$ by reflecting it across the two real segments $(-\infty, -1)$ and $(1, \infty)$. The extended domain $\hat{G}_z$ is *on a covering surface*, it has *one* sheet above the upper half-plane but *two* sheets above the lower half-plane. Its winding points are $+1$ and -1 (where $\hat{G}_z$ has the interior angle 2π), and ∞ (where $\hat{G}_z$ has the interior angle 3π).

3.2. G_z and $\hat{G}_z$ are simply connected. By the Riemann mapping theorem there exists exactly one conformal mapping $z \to \hat{z} = f(z)$ of G_z onto $\hat{G}_z$ with the three boundary values $f(-1) = -1$, $f(1) = 1$ and $f(\infty) = \infty$. (*As a complex-valued function it is not univalent, although it gives a univalent conformal mapping of G_z onto $\hat{G}_z$*, i.e. into the covering surface.) We have an analytic function in G_z with real values $f(x)$ on the real axis. By the reflection principle it can be continued to an entire function, which is here a polynomial of degree three. We obtain

$$f(z) = \frac{1}{2}(3z - z^3). \tag{2}$$

3.3. *Application: The mapping function of an L-domain formed by three squares*

We know the mapping function $w(z) : G_z \longrightarrow G_w$ of a square G_w, given by a Schwarz-Christoffel integral, whence $w'(z) = [z(1 - z^2)]^{-1/2}$. By (1) and (2), we now obtain

$$\left.\begin{aligned}
\hat{w}'(z) &= [w' \circ f(z)]f'(z) \\
&= 3\sqrt{2}[z(\sqrt{3} + z)(\sqrt{3} - z)(2 + z)(2 - z)]^{-1/2},
\end{aligned}\right\} \tag{3}$$

which gives the Schwarz-Christoffel integral $\hat{w}(z) : G_z \longrightarrow \hat{G}_w$.

We verify that $\hat{G}_w$ has six right angles, namely at $\hat{w}(\infty)$ (reentrant corner) and $\hat{w}(-2)$, $\hat{w}(-\sqrt{3})$, $\hat{w}(0)$, $\hat{w}(\sqrt{3})$, $\hat{w}(2)$ (protruding corners). We further verify that, on the upper half-plane G_z, the "quadrilateral" with distinguished boundary points $-\sqrt{3}$, 0, $\sqrt{3}$ and 2 has modulus $\sqrt{3}$, see [2], pp. 317–318, and [4], p.224. The "quadrilateral" on G_z with distinguished boundary points ∞, -2, $\sqrt{3}$ and 2 has also the same modulus $\sqrt{3}$ as it should be.

The mapping function of the L-domain $\hat{G}_w$ and the above moduli $\sqrt{3}$ have been calculated earlier by Gaier [1].

§4 Regular polygons, extension by reflection of the whole domain across all boundary segments

4.1. We know the Schwarz-Christoffel integral $w(z)$ mapping the upper half-plane G_z onto a regular polygon G_w with n sides. The extension of G_w by reflection across *all* sides gives a new polygon $\hat{G}_w$ with $n(n-1)$ sides. But, *if $n > 6$*, $\hat{G}_w$ should be considered *on a covering surface* with winding points above the n vertices of G_w.

Let $w(z_1), w(z_2), \ldots, w(z_n)$ be those vertices. We shall construct the mapping function $\hat{w}(z) : G_z \longrightarrow \hat{G}_w$ *such that* $\hat{w}(z_k) = w(z_k)$, $k = 1, 2, \ldots, n$.

4.2. *Trivial case $n = 2$. Infinite strips*

We consider the upper half-plane G_z as a "bilateral" with distinguished boundary points 0 and ∞. We extend G_z to $\hat{G}_z$ by reflection across the positive *and* across the negative real axis. Thus, $\hat{G}_z$ is the wedge $-\pi < \arg \hat{z} < 2\pi$ on the Riemann surface of the logarithmic function. Here $\hat{z} = f(z) = e^{-i\pi} z^3$. Let G_w be the strip $0 < \operatorname{Im} w < a$ and $w(z) = (a/\pi) \ln z$. Then $\hat{G}_w$ is the strip $-a < \operatorname{Im} \hat{w} < 2a$. By (1),

$$\hat{w}(z) = w \circ f(z) = (a/\pi) \ln(e^{-i\pi} z^3) = -ia + (3a/\pi) \ln z$$

which is trivial, of course.

Remark: $f(i) = i$ and $f'(i) = 3$, this is the *ratio of conformal radii*

$$\kappa_2 := \frac{R_{ia/2}(\hat{G}_w)}{R_{ia/2}(G_w)} = 3, \quad \text{of course.} \tag{4}$$

4.3. *Case $n = 3$. Equilateral triangles*

We consider the upper half-plane G_z as a "trilateral" with distinguished boundary points ∞, -1 and 1. Its *"center" (where the harmonic measure of each "side" is $1/3$)* is the point $i\sqrt{3}$. The extended domain $\hat{G}_z$ is *on a covering surface* with winding points above ∞, -1, and 1, where the interior angle is 3π. $\hat{G}_z$ has *one* sheet above the upper half-plane, *three* sheets above the lower half-plane. We impose the three boundary values $f(\infty) = \infty$, $f(-1) = -1$, $f(1) = 1$, then by the Riemann mapping theorem there exists a unique mapping function $\hat{z} = f(z) : G_z \longrightarrow \hat{G}_z$. This function is odd, it has a simple pole at 0, and the three imposed values have multiplicity 3. By the reflection principle, the function $zf(z)$ can be continued to an entire function and is therefore an even polynomial of degree 4. We obtain the rational function

$$f(z) = \frac{-z^4 + 6z^2 + 3}{8z}. \tag{5}$$

By the Schwarz-Christoffel formula, the mapping function $w(z)$ of the upper half-plane G_z onto an equilateral triangle G_w with vertices $w(\infty)$, $w(-1)$ and $w(1)$ has the derivative $w'(z) = C(z^2 - 1)^{-2/3}$. Applying (1), we obtain

$$\hat{w}'(z) = [w' \circ f(z)]f'(z) = -6C[z(z + 3)(z - 3)]^{-2/3}. \tag{6}$$

We verify that $\hat{G}_w$ is again an equilateral triangle, with vertices $\hat{w}(-3)$, $\hat{w}(0)$ and $\hat{w}(3)$.

Remark: $f(i\sqrt{3}) = i\sqrt{3}$ and $f'(i\sqrt{3}) = 2$, this is the *ratio of conformal radii*

$$\kappa_3 := \frac{R_{w_\circ}(\hat{G}_w)}{R_{w_\circ}(G_w)} = 2, \quad \text{of course,} \tag{7}$$

where $w_\circ = w(i\sqrt{3}) = \hat{w}(i\sqrt{3})$ is the center of G_w and of $\hat{G}_w$.

4.4. *Case $n = 4$. Square and Swiss cross*

We now consider the upper half-plane G_z as a *"quadrilateral"* of modulus one with the distinguished boundary points ∞, -1, 0 and 1. Its "center" is the point i. The extended domain $\hat{G}_z$ is again a quadrilateral of modulus one, now on a covering surface. It has the same four distinguished boundary points, now as winding points with interior angle 3π, and it has the same center i. $\hat{G}_z$ has *one* sheet above the upper half-plane, *four* sheets above the lower half-plane. We are interested in the mapping function $\hat{z} = f(z) : G_z \longrightarrow \hat{G}_z$ such that $f(\infty) = \infty$, $f(-1) = -1$, $f(0) = 0$ and $f(1) = 1$. We obtain

$$f(z) = \frac{-z^5 + 5z^3}{5z^2 - 1}. \tag{8}$$

By the Schwarz-Christoffel formula, the mapping function $w(z)$ of the upper half-plane onto a square G_w with vertices $w(\infty)$, $w(-1)$, $w(0)$ and $w(1)$ has the derivative $w'(z) = C[z(1 - z^2)]^{-1/2}$. $\hat{G}_w$ is now a *Swiss cross*. Applying (1), we obtain

$$\left.\begin{aligned}
\hat{w}'(z) &= [w' \circ f(z)]f'(z) \\[2mm]
&= -15C\sqrt{\frac{z(z^2 - 1)}{(5z^2 - 1)(z^2 - 5)(z^2 + 3z + 1)(z^2 - 3z + 1)}}
\end{aligned}\right\} \quad (9)$$

We see that $\hat{G}_w$ has twelve right angles: four reentrant corners $\hat{w}(\infty)$, $\hat{w}(-1)$, $\hat{w}(0)$ and $\hat{w}(1)$, eight protruding corners $\hat{w}\left(\pm\frac{3-\sqrt{5}}{2}\right)$, $\hat{w}\left(\pm\frac{1}{\sqrt{5}}\right)$, $\hat{w}(\pm\sqrt{5})$ and $\hat{w}\left(\pm\frac{3+\sqrt{5}}{2}\right)$.

Remark: $f(i) = i$ and $f'(i) = 5/3$, whence the *ratio of conformal radii*

$$\kappa_4 := \frac{R_{w_\circ}(\hat{G}_w)}{R_{w_\circ}(G_w)} = \frac{5}{3}, \quad (10)$$

where $w_\circ = w(i) = \hat{w}(i)$ is the center both of the square G_w and of the Swiss cross $\hat{G}_w$.

— Of course, the same method applies if G_w is a rhombus rather than a square.

4.5. *A conjecture*

The values obtained in (4), (7) and (10) for ratios of conformal radii:

$$\kappa_2 = 3, \qquad \kappa_3 = 2, \qquad \kappa_4 = 5/3$$

suggest the *conjecture* that, for all $n \geq 2$,

$$\kappa_n \overset{?}{=} \frac{n+1}{n-1} \quad conj. \quad (11)$$

This is also true for $n = 6$, *since a similar calculation yields* $\kappa_6 = 7/5$.

Remarks

(a) Each value of κ_n corresponds to a particular value of Green's function, see [3], p. 148 (case $n = 3$).

(b) The simple idea expressed in Section 1 has been applied in [6] to symmetric domains, for example regular polygons, which are extended by *"sector reflections"*, whereas here we always reflect the *whole* domain. In the case of "sector reflections" some ratios of conformal radii had already been determined in [5].

(c) Since we write w for the mapping $G_z \longrightarrow G_w$ as well as for $\hat{G}_z \longrightarrow \hat{G}_w$, it should be clear that, in (1), $w(z)$ means the *global* function, rather than a particular expression of it in G_z.

(d) These methods may be known since over a century, the author would be thankful for any bibliographical indications.

References

[1] D. Gaier, *Ermittlung des konformen Moduls von Vierecken mit Differenzenmethoden.* Numer. Math. *19*, 179–194 (1972).

[2] J. Hersch, *Longueurs extrémales et théorie des fonctions.* Comment. Math. Helv. *29*, 301–337 (1955).

[3] J. Hersch, *Erweiterte Symmetrieeigenschaften von Lösungen gewisser linearer Rand- und Eigenwertprobleme.* J. Reine Angew. Math. *218*, 143–158 (1965).

[4] J. Hersch, *On harmonic measures, conformal moduli and some elementary symmetry methods.* J. d'Analyse Math. *42*, 211–228 (1982/83).

[5] J. Hersch, *On the reflection principle and some elementary ratios of conformal radii.* J. d'Analyse Math. *44*, 251–268 (1984/85).

[6] J. Hersch, *On the mapping functions of domains extended by sector reflections.* Complex Variables *9*, 199–209 (1987).

Summary

If we know a function $w(z)$ mapping the upper half-plane G_z conformally onto a polygon G_w, we can use the reflection principle to construct a mapping $\hat{w}(z)$ of G_z onto a larger polygon $\hat{G}_w$, obtained by reflecting all of G_w across one or several sides.

Mathematics
ETH-Zentrum
CH–8092 Zürich
Switzerland

Alfred Huber

Konforme Verheftung
und logarithmisches Potential

1. Das Problem der konformen Verheftung gehört in das Gebiet der geometrischen Funktionentheorie. Es überrascht daher keineswegs, dass es mit der Theorie des logarithmischen Potentials in Verbindung gebracht werden kann. Schon 1960 hat H. Grunsky [1] auf einen solchen Zusammenhang hingewiesen. In der vorliegenden Note berichten wir kurz über einen andern von der konformen Verheftung zum logarithmischen Potential führenden Weg.

Gegeben sei eine orientierungstreue und homöomorphe Abbildung,

$$\Phi \; : \; e^{i\theta} \to e^{i\varphi(\theta)} \quad (\theta, \varphi \in I\!R), \tag{1.1}$$

des Einheitskreises $C = \{z \in \mathbb{C} \mid |z| = 1\}$ auf sich. Die Funktion Φ sei (in einer Umgebung von C definiert und) analytisch, und es sei $\Phi' \neq 0$ auf C. Verheftet man das Innere von C mit dem Äusseren durch Identifikation von z mit $\Phi(z)$ für alle $z \in C$, so entsteht eine Riemannsche Fläche. Nach dem Uniformisierungssatz gibt es eine analytische Jordankurve Γ (im folgenden "Verheftungskurve" genannt), eine konforme Abbildung F des Inneren von C auf das Innere von Γ und eine konforme Abbildung G des Äusseren von C auf das Äussere von Γ derart, dass

$$F(e^{i\theta}) = G(\Phi(e^{i\theta})) \quad \text{für alle} \quad \theta \in I\!R. \tag{1.2}$$

In (1.2) sind die Werte von F und G auf dem Rand C einzusetzen. Diese existieren, denn da Γ analytisch ist, besitzen F und G nach dem Spiegelungsprinzip sogar eine analytische Fortsetzung (mit nichtverschwindender Ableitung) über den Gebietsrand C hinaus. Auf C gilt $\Phi = G^{-1} \circ F$, so dass wir schliessen können: *Hat ein (a priori beliebiger) orientierungstreuer Homöomorphismus Φ von C auf sich die Eigenschaft, dass er eine konforme*

Verheftung mit analytischer Verheftungskurve Γ erzeugt, so erfüllt er sämtliche eingangs erwähnten Voraussetzungen. Normiert man $G(\infty) = \infty$, so ist Γ bis auf eine Ähnlichkeitstransformation festgelegt.

Diese Voraussetzungen sind um einiges schärfer, als dies für unsere Zwecke unbedingt notwendig wäre. So ist z.B. die Existenz einer solchen — dann natürlich nicht mehr analytischen — Verheftungskurve Γ unter weit schwächeren Annahmen über Φ gewährleistet, wie dies von A. Pfluger [6] und von O. Lehto und K.I. Virtanen [5] gezeigt wurde. Der Kürze und der Bequemlichkeit halber wurden jedoch die vorliegenden engen Voraussetzungen gewählt.

2. Die Menge E aller harmonischen Funktionen, deren Definitionsgebiet die Kreisscheibe $\{z \mid |z| \leq 1\}$ enthält, teilen wir in Klassen ein: zwei solche Funktionen sollen genau dann derselben Klasse angehören, falls sie sich nur um eine Konstante unterscheiden, d.h. falls sie Lösungen desselben Neumannproblems für den Einheitskreis sind.

Definition. Jeder solchen Klasse K wird eine Klasse K_o zugeordnet. Diese ist dadurch charakterisiert, dass für $h(e^{i\theta}) \in K$ und $h_o(e^{i\varphi}) \in K_o$ die Beziehung

$$\varphi'(\theta)\frac{\partial h_o}{\partial n}(e^{i\varphi(\theta)}) = \frac{\partial h}{\partial n}(e^{i\theta}) \tag{2.1}$$

für alle reellen θ erfüllt ist. (Mit n wird hier und im folgenden die Richtung der äusseren Normalen von C bezeichnet.)

Anmerkung. Um einzusehen, dass es zu jeder Klasse $K \subset E$ eine zugeordnete Klasse $K_o \subset E$ gibt, betrachte man irgendeinen Repräsentanten $h \in K$. Sei $\tilde{h}$ eine zu h konjugiert harmonische Funktion. Durch Lösung des entsprechenden Dirichletproblems erhält man eine im Innern von C harmonische Funktion $\tilde{h}_o$ mit stetigen Randwerten auf C, welche für alle $\theta \in I\!\!R$ die Bedingung

$$\tilde{h}_o(\Phi(e^{i\theta})) = \tilde{h}(e^{i\theta}) \tag{2.2}$$

erfüllen. Die Funktion $\tilde{h}_o \circ \Phi$ ist harmonisch in einem Kreisring mit äusserem Rand C und kann — dies folgt aus (2.2) unter Anwendung des Spiegelungsprinzips — über C hinaus harmonisch fortgesetzt werden. Da $\tilde{h}_o \circ \Phi$ in einer vollen Umgebung von C harmonisch definiert werden kann, gilt dasselbe auch von $\tilde{h}_o = (\tilde{h}_o \circ \Phi) \circ \Phi^{-1}$. Bezeichnen wir nun mit $-h_o$ eine zu $\tilde{h}_o$ konjugiert harmonische Funktion — die natürlich ebenfalls über C hinaus harmonisch fortsetzbar ist — , so liegt h_o in K_o, da die Relation (2.1) durch (2.2) impliziert wird.

Der Note [3] kann das folgende Resultat entnommen werden:

Satz 1. *Die Abbildungsfunktionen F und G erfüllen die Gleichung*

$$\left.\begin{array}{c}
\displaystyle\int_0^{2\pi} h(e^{i\theta})\frac{d}{d\theta}\arg F'(e^{i\theta})d\theta + \int_0^{2\pi} h_\circ(e^{i\varphi})\frac{d}{d\varphi}\arg G'(e^{i\varphi})d\varphi \\[4mm]
= \displaystyle\int_0^{2\pi} h(e^{i\theta})\frac{\partial U}{\partial n}(e^{i\theta})d\theta
\end{array}\right\} \qquad (2.3)$$

für jedes Paar (gemäss obiger Definition) zugeordneter harmonischer Funktionen $H \in K$ und $h_\circ \in K_\circ$ aus E. Dabei bezeichnen U die Lösung des Dirichletschen Problems für das Innere von C mit Randwerten $\log \varphi'(\theta)$ auf C, und $\partial/\partial n$ die Ableitung in Richtung der äusseren Normalen von C.

Anmerkung. Beim Vergleich von Satz 1 mit Theorem 2 in [3] wird man bemerken, dass ein negatives Vorzeichen in ein positives umgewandelt wurde. Es handelt sich dabei um die Korrektur eines Vorzeichenfehlers, der sich in Herleitung und Formulierung von Theorem 2 eingeschlichen hatte.

3. Bei der Anwendung von Satz 1 erscheint es zweckmässig, von dem durch die Normalableitung von h (bzw. von $h_\circ$) erzeugten Mass auszugehen. Nach (2.1) wird dieses nämlich bei der Abbildung Φ verpflanzt

$$\frac{\partial h}{\partial n}(e^{i\theta})d\theta = \frac{\partial h_\circ}{\partial n}(e^{i\varphi})d\varphi. \qquad (3.1)$$

Bei regulärem Randverhalten — wie in unserem Fall — ergeben sich daraus die Randwerte $h(e^{i\theta})$ (bzw. $h_\circ(e^{i\varphi})$) nach einer bekannten Formel von Dini (vgl. z.B. [4], Kap. III, §1, Abschnitt 44, Formel (21)),

$$h(e^{i\theta}) = -\frac{1}{\pi}\int_0^{2\pi} \frac{\partial h}{\partial n}(e^{it})\log|e^{i\theta} - e^{it}|dt + \text{const.} \qquad (3.2)$$

4. Die Funktionen h und $h_\circ$, die wir nun in (2.3) einsetzen, legen wir also durch Vorgabe ihrer Normalableitungen fest. Sie hängen von einem Parameter z ab, der über eine Umgebung $\Omega = \{z \in \mathbb{C} \mid 1 - \epsilon < |z| < 1 + \epsilon\}$, $\epsilon > 0$, des Kreises C variieren darf. Diese sei so klein, dass Φ in ihr analytisch und $\Phi' \neq 0$ ist. Für alle $\theta \in R$ soll gelten

$$\frac{\partial h}{\partial n}(e^{i\theta}) = \log|z - e^{i\theta}| + A(|z|). \qquad (4.1)$$

Die Zahl $A(|z|)$ wird dadurch festgelegt, dass $\int_0^{2\pi}(\partial h/\partial n)(e^{i\theta})d\theta = 0$. Aus (4.1) und (2.1) folgt

$$\frac{\partial h_{\mathrm{o}}}{\partial n}(e^{i\varphi}) = [\log|z - \Phi^{-1}(e^{i\varphi})| + A(|z|)]\frac{d\theta}{d\varphi}. \tag{4.2}$$

Die Bedingung $\int_0^{2\pi}(\partial h_{\mathrm{o}}/\partial n)(e^{i\varphi})d\varphi = 0$ ist dann ebenfalls erfüllt.

Unter Anwendung der Dinischen Formel (3.2) folgt für alle $z \in \Omega$ aus (2.3), (4.1) und (4.2)

$$I(z) - K(z) + L(z) = 0, \tag{4.3}$$

wobei

$$
\begin{aligned}
I(z) &= \int_0^{2\pi}\frac{d}{d\theta}\arg F'(e^{i\theta})\int_0^{2\pi}(\log|e^{i\theta} - e^{it}|)\times \\
&\qquad \times (\log|z - e^{it}| + A(|z|))dt\,d\theta \\
&= \int_0^{2\pi}\frac{d}{d\theta}\arg F'(e^{i\theta})\int_0^{2\pi}(\log|e^{i\theta} - e^{it}|)(\log|z - e^{it}|)dt\,d\theta,
\end{aligned}
$$

$$
K(z) = \int_0^{2\pi}\frac{\partial U}{\partial n}(e^{i\theta})\int_0^{2\pi}(\log|e^{i\theta} - e^{it}|)(\log|z - e^{it}|)dt\,d\theta,
$$

$$
\begin{aligned}
L(z) &= \int_0^{2\pi}\frac{d}{d\varphi}\arg G'(e^{i\varphi})\int_0^{2\pi}(\log|e^{it} - e^{i\varphi}|)\times \\
&\qquad \times (\log|z - \Phi^{-1}(e^{it})| + A(|z|))\frac{d\theta}{d\varphi}(e^{it})dt\,d\varphi \\
&= \int_0^{2\pi}\frac{d}{d\varphi}\arg G'(e^{i\varphi})\int_0^{2\pi}(\log|\Phi(e^{is}) - e^{i\varphi}|)\times \\
&\qquad \times (\log|z - e^{is}| + A(|z|))ds\,d\varphi \\
&= \int_0^{2\pi}\frac{d}{d\varphi}\arg G'(e^{i\varphi})\int_0^{2\pi}(\log|\Phi(e^{is}) - e^{i\varphi}|)(\log|z - e^{is}|)ds\,d\varphi.
\end{aligned}
$$

Bemerkungen.

1) Es wurde hier davon Gebrauch gemacht, dass

$$\int_0^{2\pi} \frac{d}{d\theta} \arg F'(e^{i\theta})d\theta = 0, \tag{4.4}$$

$$\int_0^{2\pi} \frac{d}{d\varphi} \arg G'(e^{i\varphi})d\varphi = 0, \tag{4.5}$$

$$\int_0^{2\pi} \frac{\partial U}{\partial n}(e^{i\theta})d\theta = 0. \tag{4.6}$$

2) Im Fall $|z| = 1$ kann Satz 1 nicht ohne weiteres angewandt werden, da die Funktionen h und $h_\circ$ auf C eine Singularität aufweisen. Man kann sich hier durch einen Grenzübergang behelfen oder ganz einfach bemerken, dass die Werte der Funktion $z \to I(z) - K(z) + L(z)$ auf der zweidimensionalen Lebesgueschen Nullmenge C für die Definition der von ihr erzeugten Distribution ohne Einfluss sind.

Nun wird auf die (fast überall) in Ω gültige Relation (4.3) der (zweidimensionale, im Sinne der Theorie der Distributionen aufzufassende) Laplaceoperator Δ_z angewandt. Dabei erhalten wir: $\Delta_z I$ ist ein Mass auf C mit der Dichtefunktion

$$2\pi \int_0^{2\pi} (\log |e^{ix} - e^{i\theta}|)\frac{d}{d\theta} \arg F'(e^{i\theta})d\theta, \tag{4.7}$$

$0 \leqq x \leqq 2\pi$; $\Delta_z K$ ist ein Mass auf C mit der Dichtefunktion

$$2\pi \int_0^{2\pi} (\log |e^{ix} - e^{i\theta}|)\frac{\partial U}{\partial n}(e^{i\theta})d\theta, \tag{4.8}$$

$0 \leqq x \leqq 2\pi$; $\Delta_z L$ ist ein Mass auf C mit der Dichtefunktion

$$2\pi \int_0^{2\pi} (\log |\Phi(e^{ix}) - e^{i\varphi}|)\frac{d}{d\varphi} \arg G'(e^{i\varphi})d\varphi, \tag{4.9}$$

$0 \leqq x \leqq 2\pi$.

Aus (4.3), (4.7), (4.8) und (4.9) ergibt sich eine für Diracsche Masse gültige Relation, deren Erweiterung auf allgemeinere Masse keinerlei Schwierigkeiten bietet:

Satz 2. *Sei μ ein Borelsches Mass auf C, und sei μ^* die Verpflanzung von μ bei der Abbildung Φ. Dann gilt*

$$\left. \begin{aligned} &\int\limits_{C} d\mu(x) \int\limits_{0}^{2\pi} \left(\log\left|e^{ix} - e^{i\theta}\right|\right) \frac{d}{d\theta} \arg F'(e^{i\theta}) d\theta + \\ &+ \int\limits_{C} d\mu^*(y) \int\limits_{0}^{2\pi} \left(\log\left|e^{iy} - e^{i\varphi}\right|\right) \frac{d}{d\varphi} \arg G'(e^{i\varphi}) d\varphi = \\ &= \int\limits_{C} d\mu(x) \int\limits_{0}^{2\pi} \left(\log\left|e^{ix} - e^{i\theta}\right|\right) \frac{\partial U}{\partial n}(e^{i\theta}) d\theta. \end{aligned} \right\} \qquad (4.10)$$

Auch die (von innen und von aussen betrachtete) Tangentendrehung der Kurve Γ ist ein Mass auf C, das bei der Abbildung Φ verpflanzt wird. Es gilt nämlich

$$\left(1 + \frac{d}{d\theta} \arg F'(e^{i\theta})\right) d\theta = \left(1 + \frac{d}{d\varphi} \arg G'(e^{i\varphi})\right) d\varphi.$$

Um dieses Mass besser hervortreten zu lassen, geben wir deshalb noch eine andere — wegen (4.4) und (4.5) zu (4.10) äquivalente — Formulierung desselben Resultates:

Satz 3. *Sei μ ein Borelsches Mass auf C mit $\mu(C) = 0$, und sei μ^* die Verpflanzung von μ bei der Abbildung Φ. Dann gilt*

$$\left. \begin{aligned} &\int\limits_{C} d\mu(x) \int\limits_{0}^{2\pi} \left(\log\left|e^{ix} - e^{i\theta}\right|\right) \left[1 + \frac{d}{d\theta} \arg F'(e^{i\theta})\right] d\theta + \\ &+ \int\limits_{C} d\mu^*(y) \int\limits_{0}^{2\pi} \left(\log\left|e^{iy} - e^{i\varphi}\right|\right) \left[1 + \frac{d}{d\varphi} \arg G'(e^{i\varphi})\right] d\varphi = \\ &= \int\limits_{C} d\mu(x) \int\limits_{0}^{2\pi} \left(\log\left|e^{ix} - e^{i\theta}\right|\right) \frac{\partial U}{\partial n}(e^{i\theta}) d\theta. \end{aligned} \right\} \qquad (4.11)$$

Um mit Hilfe von (4.11) zu konkreten geometrischen Aussagen über die
Verheftungskurve Γ zu gelangen, wird man etwas über das Verhalten von
Φ (z.B. über die Variationsbreite des Quotienten $|e^{i\varphi(x)} - e^{i\varphi(\theta)}|/|e^{ix} - e^{i\theta}|$)
voraussetzen müssen. Der potentialtheoretische Aspekt der linken Seite von
(4.11) legt die Vermutung nahe, der hier ins Auge gefasste Ansatz könnte
seine natürliche Grenze bei den Verheftungskurven beschränkten Drehung
finden.

Literaturverzeichnis

[1] H. Grunsky: Eine Grundaufgabe der Uniformisierungstheorie als Extre-
malproblem, Math. Ann. *139* (1960), 204–216.

[2] A. Huber: Konforme Verheftung und Dirichletsches Prinzip, Ann. Acad.
Sci. Fenn., Series A.I. Mathematica, *10* (1985) (O. Lehto gewidmet),
261–265.

[3] A. Huber: Change of Angles in Conformal Welding, Complex Variables
7 (1986) (H. Grunsky gewidmet), 79–82.

[4] M.A. Lawrentjew und B.W. Schabat: Methoden der komplexen Funktionentheorie (Übersetzt aus dem Russischen) Deutscher Verlag der Wis-
senschaften 1967.

[5] O. Lehto und K.I. Virtanen: Quasikonforme Abbildungen, Springer-Verlag Berlin-Heidelberg-New York, 1965.

[6] A. Pfluger: Über die Konstruktion Riemannscher Flächen durch Verheftung, J. Indian Math. Soc. (N.S.) *24*, 1960, 401–412.

Mathematik
ETH-Zentrum
CH-8092 Zürich
Schweiz

| Complex | Edited by | Birkhäuser Verlag |
| Analysis | J. Hersch and A. Huber | Basel 1988 |

James A. Jenkins

On Boundary Correspondence for Domains on the Sphere

It is well known that a conformal mapping between domains bounded by Jordan curves can be extended to a homeomorphism of the closures. For many applications what is needed however is a local version of this result. In the present paper such a result is provided in a generalized context.

AMS (MOS): 30D40, 30C35.

1. The method of the extremal metric had its first primitive origins in the theory of boundary correspondence. It should not be surprising then that the most satisfactory treatment of boundary correspondence is by that method. An exposition along these lines dates from the early days of the development of the method of the extremal metric in its abstracted version and was presented by Ahlfors in his lectures at Harvard University in the spring 1947. This however contained a serious gap in the proof of the transitivity of equivalence for fundamental sequences. Several later presentations [4; 1, p.57] undertook to overcome this difficulty by artificial definitions. This approach is particularly effective in proving that a conformal mapping between Jordan domains can be extended to a homeomorphism between the closures. A later result by Arsove [2] shows that a conformal mapping from a Jordan domain to a domain whose boundary can be continuously parametrized as a curve can be extended to be continuous on the closure.

Actually these latter results are properly of local character in a sense to be made clear hereafter although all published proofs have a distinct global character. This does not deter some authors from applying the global result without further justification to draw local conclusions. In the present paper we will consider the results in the proper local form. Since no completely satisfactory treatment of this material appears in print we will develop the exposition from first principles.

2. Let D be a domain on the sphere which has a free continuum boundary component C. By this we mean that no point of C is an accumulation point of boundary points of D not on C. Let Δ be the simply-connected domain complementary to C which contains D. Let E be a continuum contained in Δ which contains all other boundary components of D (and naturally is arbitrary if D is itself simply-connected).

Definition 1. By a crosscut of Δ we mean an open arc in Δ which tends to C in each sense.

Definition 2. By a fundamental sequence associated with C and D we mean a non-increasing sequence of non-void sets $\{E_n\}$, $E_n \subset \Delta - E$, tending to C and such that if Γ_n denotes the family of locally rectifiable crosscuts of Δ separating E and E_n then for the module $m(\Gamma_n)$ of Γ_n

$$\lim_{n \to \infty} m(\Gamma_n) = \infty.$$

Definition 3. By a C-fundamental sequence (fundamental in the sense of Carathéodory) associated with C and D we mean a non-increasing sequence of non-void sets $\{E_n\}$, $E_n \subset \Delta - E$, tending to C and such that given $\epsilon > 0$ for $n \geq N(\epsilon)$ there exists a crosscut of Δ in $\Delta - E$ separating E and E_n of spherical diameter $< \epsilon$.

Theorem 1. *A sequence $\{E_n\}$ is C-fundamental if and only if it is fundamental.*

We first prove necessity. Let δ be the spherical distance from E to C. Given ϵ, $0 < \epsilon < \frac{1}{2}\delta$ let γ be a crosscut of Δ in $\Delta - E$ of spherical diameter $< \epsilon$ separating E and E_n for $n \geq N(\epsilon)$. About a point of γ we draw spherical circles of radii ϵ and $\frac{1}{2}\delta$. In the circular ring R bounded by them every concentric circle c_r of spherical radius r contains a crosscut γ_r of Δ separating E and E_n. Let $\rho|dz|$ be an admissible metric in $\Delta - E$ for the module problem defining $m(\Gamma_n)$ in L-normalization [3, p.14]. If we set

$$\rho^* = \rho \text{ in } R \cap \Delta$$

$$\rho^* = 0 \text{ elsewhere in } R$$

then $\int_{c_r} \rho^*|dz| \geq 1$. Thus $\int_R \rho^{*2}dA$ is at least equal to the module of the ring R (see the proof in [3, p.18]), namely $\frac{1}{2\pi} \log \frac{\delta}{2\epsilon}(1 - \epsilon^2)^{1/2}(1 - \frac{1}{4}\delta^2)^{-1/2}$, and so

$$m(\Gamma_n) \geq \frac{1}{2\pi} \log \frac{\delta}{2\epsilon}(1 - \epsilon^2)^{1/2}(1 - \frac{1}{4}\delta^2)^{-1/2}.$$

Thus $\lim\limits_{n \to \infty} m(\Gamma_n) = \infty$ and $\{E_n\}$ is fundamental.

To prove sufficiency we consider the module problem for $m(\Gamma_n)$ in the general form [3, Definition 2·3] and use for a comparison metric the spherical metric $\sigma|dz|$. Since $\int\limits_{\Delta - E} \sigma^2 dA$ is finite, say $= A$, we have

$$\operatornamewithlimits{g.l.b}_{\gamma \in \Gamma_n} \int_\gamma \sigma|dz| < \left(\frac{A}{(m(\Gamma_n))^2}\right)^{1/2}$$

and as soon as $m(\Gamma_n) > A\epsilon^{-2}$ we have a crosscut in Γ_n of spherical length and thus spherical diameter $< \epsilon$ and $\{E_n\}$ is $\mathcal{C}$-fundamental.

Definition 4. Two fundamental sequences $\{E_n\}$ and $\{E'_n\}$ are said to be equivalent if $\{E_n \cup E'_n\}$ is also a fundamental sequence.

This relation is evidently reflexive and symmetric. To see that it is transitive we suppose $\{E_n\}$ and $\{E'_n\}$ are equivalent as also $\{E'_n\}$ and $\{E''_n\}$ so that for n sufficiently large there will be crosscuts γ, γ' of Δ of spherical diameter less than a prescribed $\epsilon > 0$ separating $E_n \cup E'_n$ and $E'_n \cup E''_n$ from E. We distinguish two cases. First if $\gamma \cap \gamma' = 0$ one crosscut will separate the other from E and thus $E_n \cup E''_n$, for n sufficiently large, from E. If $\gamma \cap \gamma' \neq 0$ the argument of the first part of Theorem 1 shows that an open arc on a circle c_r can be chosen to have diameter tending to zero with ϵ and to separate $E_n \cup E''_n$, for n sufficiently large, from E.

Definition 5. Equivalence classes of fundamental sequences are called prime ends or boundary elements associated with C and D.

Finally we remark that while these definitions may appear to depend on the choice of E the entities in question essentially do not in the sense that terminal sequences will retain the property of being fundamental, as is seen by an easy argument.

3. We are now ready to state our principal result.

Theorem 2. *Let D be a domain on the sphere, C a free continuum boundary component of D. Let H be a closed subset of C which can be parametrized as a continuous path and such that there exists a simply-connected subdomain Ξ of D whose boundary consists of H and L with $L \subset D$. Let D be the conformal image under f of a domain G so that C corresponds to a Jordan curve J. Then there exists an open subarc j on J and a continuous extension of f to j such that $f(j)$ contains all points of H except perhaps those of $H \cap ClL$.*

We begin by identifying prime ends associated with C and D corresponding to points of H. Let P be a point of $H - ClL$ and let $r > 0$ be so small that $c(P, r) = \{d(P, Q) = r\}$ does not meet $E \cup (\partial D - H)$ (here $d(P, Q)$ denotes the spherical distance between P and Q). $c(P, r) \cap D$ consists of an at most countable set of crosscuts of D denoted generically by η. They are partially ordered by the relation $\eta_2 > \eta_1$ which means η_1 separates η_2 from E. It is readily verified that every linearly ordered subset contains only finitely many elements and thus a maximal element. Consider now such maximal elements for a sequence of values $\{r_j\}$ tending monotonically to zero and linearly ordered (proper) sequences $\{\eta_j\}_1^\infty$ of these elements. There is at least one such sequence. Corresponding to them and the associated sets $E(\eta_j)$ (which is the subset of $D - \eta_j$ not containing E) we have a number of properties.

(i) $\{E(\eta_j)\}$ is a fundamental sequence.

(ii) $\{E(\eta_j)\}$ converges to the given point P.

Indeed it is readily verified that the diameter of $E(\eta_j)$ tends to zero as j tends to infinity. This uses the continuous parametrization of H.

(iii) If $\{\eta_j\}$, $\{\eta_j'\}$ are sequences with $\eta_k \neq \eta_k'$ then $\eta_l \neq \eta_l'$ for $l > k$. This uses the existence of Ξ.

(iv) For each sequence $\{\eta_j\}$ there is an open arc $\lambda\{\eta_j\}$ in $D - E$ with limiting end points on E and at the given point P constructed of arcs in $ClE(\eta_j) - E(\eta_{j+1})$ joining successively points on η_j and η_{j+1}. Let $\lambda_l\{\eta_j\}$ denote the closed arc from the end point on E to the point on η_l.

(v) No two distinct $\{E(\eta_j)\}$ are equivalent. If $\lambda_l\{\eta_j\}$ has distance δ from C and $E(\eta_t)$ has diameter $\delta^* < \delta$ then a crosscut γ of Δ of diameter $< \delta^*$ separating some $E(\eta_s)$ from E must lie in $E(\eta_m)$ if $r_m > \delta^* + r_t$ because $Cl\gamma$ must meet $Cl\lambda\{\eta_j\} - \lambda_t\{\eta_j\}$ and thus cannot meet η_m.

In proving Theorem 2 we may assume first that J is a circumference. The above properties show that the prime ends associated with J and G are in $(1,1)$ correspondence with the points of J. Indeed if $T \in J$ and $\{E_n(T)\}$ is a corresponding fundamental sequence constructed as above then if $\{E_n\}$ is a fundamental sequence for J and G and $T \in \cap ClE_n$ the E_n converge to T and $\{E_n\}$ is equivalent to $\{E_n(T)\}$. For $P \in H - ClL$ $\{\lambda\{\eta_j\} - \lambda_n\{\eta_j\}\}$ is a fundamental sequence thus $f^{-1}(\lambda)$ has a limiting end point T on J. Distinct points P lead to distinct points T. We define $f(T) = P$. $f^{-1}(\Xi)$ is a simply-connected subdomain of the complementary domain $I(J)$ of J containing G. Its boundary consists of points in $I(J)$ corresponding to L and points on J. Let T be one of the latter points at positive distance from $f^{-1}(L)$ and suppose the sets in $\{E_n(T)\}$ lie in $f^{-1}(\Xi)$. Then $\{f(E_n(T))\}$ is a fundamental sequence of sets in Ξ and $\bigcap_{n=1}^{\infty} Clf(E_n(T))$ contains a point P of $H - ClL$. As in the proof of (v) and using the same notation a crosscut γ of Δ of diameter $< \delta^*$ separating some $f(E_n(T))$ from E will lie in $E(\eta_m)$ for suitable m, thus

$\{E_n(T)\}$ is equivalent to the fundamental sequence determined as above by $\lambda\{\eta_j\}$. Since a subarc of J joining two points such as T lies in $\partial f^{-1}(\Xi)$ all such points make up an open subarc j of J to which f is extended so that $H \supset f(j) \supset H - ClL$. The extended mapping is continuous at points of j for approach in $I(J)$ and since each $f^{-1}(\eta_j)$ is a crosscut of $I(J)$ with both end points on j it is also continuous for approach along j.

Corollary 1. *A conformal mapping of a domain with a free circumference boundary component onto a domain with a corresponding free Jordan curve boundary component J can be extended to these components to be a homeomorphism.*

We need apply the above result only to each of two arcs covering J and overlapping at their ends.

Now in the proof of Theorem 2 we can drop the assumption that J is a circumference thus completing the proof in general.

Corollary 2. *A conformal mapping of a domain with a free Jordan curve boundary component onto a domain whose corresponding boundary component can be parametrized as a continuous path can be extended to map the former continuously onto the latter.*

References

[1] L.V. Ahlfors, *Conformal Invariants*, McGraw-Hill, New York, 1973.

[2] M.G. Arsove, The Osgood-Taylor-Carathéodory theorem, *Proc. A.M.S.* 19 (1968), 38–44.

[3] James A. Jenkins, *Univalent Functions and Conformal Mapping*, Springer-Verlag, Berlin-Göttingen-Heidelberg, 1958.

[4] E.C. Schlesinger, Conformal invariants and prime ends, *Amer. J. Math.* 80 (1958), 83–102.

Research supported in part by the National Science Foundation. A substantial amount of work on this paper was done while the author was visiting at Science University of Tokyo at the invitation of Mitsuru Ozawa.

Department of Mathematics
Washington University
St. Louis, Missouri 63130
U.S.A.

Wilfred Kaplan

On Circulants

§1 Introduction

Circulants were introduced as determinants of matrices of the form

$$
Q = \begin{bmatrix}
c_0 & c_1 & \cdot & \cdot & \cdot & c_{n-1} \\
c_{n-1} & c_0 & c_1 & \cdot & \cdot & c_{n-2} \\
\cdot & \cdot & \cdot & \cdot & \cdot & \cdot \\
\cdot & \cdot & \cdot & \cdot & \cdot & \cdot \\
c_1 & c_2 & \cdot & \cdot & \cdot & c_0
\end{bmatrix} = (q_{ij})
\tag{1}
$$

Thus $q_{ij} = c_{\mathrm{mod}(j-i,n)}$. One now refers to *circulant matrices*. Closely related to Q is the *skew-circulant* matrix Q' in which each entry in Q above the principal diagonal is multiplied by -1.

The present paper concerns some general formulas for the determinants of such matrices and their relations to the theory of symmetric functions.

We write $p(z) = c_0 + c_1 z + \cdots + c_{n-1} z^{n-1}$ and call $p(z)$ the *polynomial associated to* Q or Q'. We write

$$
\omega = e^{2\pi i/n}, \quad z_j = \omega^{j-1}, \quad z'_j = z_j e^{\pi i/n} \quad \text{for} \quad j = 1, \ldots, n.
$$

It is well known that

$$
\det Q = \prod_{j=1}^{n} p(z_j),
\tag{2}
$$

$$
\det Q' = \prod_{j=1}^{n} p(z'_j).
\tag{3}
$$

We denote by

$$\sum a_1^{k_1} \cdots a_s^{k_s} \tag{4}$$

the standard symmetric function and denote by

$$G_n(k_1, \ldots, k_n), \quad G_n'(k_1, \ldots, k_n) \tag{5}$$

the value of (4) when $s = n$ and, for $j = 1, \ldots, n$, $a_j = z_j$ or $a_j = z_j'$ respectively. We also write

$$\left. \begin{aligned} H_n(\alpha_0, \alpha_1, \ldots, \alpha_l) &= G_n(k_1, \ldots, k_n), \\ H_n'(\alpha_0, \alpha_1, \ldots, \alpha_l) &= G_n'(k_1, \ldots, k_n) \end{aligned} \right\} \tag{6}$$

where all α_j are non-negative, $\alpha_0 + \alpha_1 + \cdots + \alpha_l = n$ and α_0 of the k_i are 0, α_1 of the k_i are $1, \ldots$.

Classical work of Faà de Bruno [2] expresses the function (4) in terms of the power sums $\sum a_1^p$ by explicit formulas. This work is used by Ore [3] to give explicit formulas for the H_n; a slight modification of Ore's results gives formulas for the H_n'.

§2   General formulas for $\det Q$ and $\det Q'$.

Theorem 1. *In the above notations,*

$$\det Q = \sum_{(\alpha)} c_0^{\alpha_0} c_1^{\alpha_1} \cdots c_{n-1}^{\alpha_{n-1}} H_n(\alpha_0, \ldots, \alpha_{n-1}), \tag{7}$$

$$\det Q' = \sum_{(\alpha)} c_0^{\alpha_0} c_1^{\alpha_1} \cdots c_{n-1}^{\alpha_{n-1}} H_n'(\alpha_0, \ldots, \alpha_{n-1}), \tag{7'}$$

where in each case the sum is over all n-tuples $(\alpha) = (\alpha_0, \alpha_1, \ldots, \alpha_{n-1})$ of integers

$$\left. \begin{aligned} &\alpha_j \geq 0, \ j = 1, \ldots, n; \quad \sum_j \alpha_j = n; \\ &\sum_j j\alpha_j \quad \text{divisible by } n. \end{aligned} \right\} \tag{8}$$

Proof. The result (7) is derived by Ore [3]; see also [4], pp. 453–454. We rederive it in a slightly different way.

Equation (2) can be written $\det Q = \Psi(1)$, where

$$\Psi(z) = \prod_{j=1}^{n} p(z_j z). \tag{9}$$

From the definition of $p(z)$

$$\Psi(z) = \sum_{k_1=0}^{n-1} \cdots \sum_{k_n=0}^{n-1} c_{k_1} \cdots c_{k_n} z_1^{k_1} \cdots z_n^{k_n} z^{k_1+\cdots+k_n}. \tag{10}$$

In (9) Ψ is a polynomial (of degree at most $n(n-1)$) and, since $\Psi(z) = \Psi(\omega z)$, Ψ is a function of z^n. Hence in (10) the coefficient of z^s is 0 except for s divisible by n. We can write

$$\Psi(z) = c_0^n + \sum_{m=1}^{n-1} d_m z^{mn}.$$

Here d_m is the sum of all quantities

$$c_{k_1} \cdots c_{k_n} z_1^{k_1} \cdots z_n^{k_n}$$

for which $k_1 + \cdots + k_n = mn$. Each such term can be written as

$$c_0^{\alpha_0} c_1^{\alpha_1} \cdots c_{n-1}^{\alpha_{n-1}} z_1^{k_1} \cdots z_n^{k_n},$$

where α_0 of the k_i are 0, α_1 of the k_i are 1,... and (α) satisfies (8). The corresponding sum in $\Psi(1)$ is

$$c_0^{\alpha_0} c_1^{\alpha_1} \cdots c_{n-1}^{\alpha_{n-1}} H_n(\alpha_0, \alpha_1, \ldots, \alpha_{n-1}).$$

This gives (7). The proof of $(7')$ is the same, with z_j' replacing z_j. $\square$

Example 1. Let $n = 3$. The set of permissible (α) is $(3, 0, 0)$, $(0, 3, 0)$, $(0, 0, 3)$, $(1, 1, 1)$ and

$$H_3(3,0,0) = \sum z_1^0 z_2^0 z_3^0 = 1, \quad H_3(0,3,0) = \sum z_1 z_2 z_3 = 1,$$

$$H_3(0,0,3) = \sum z_1^2 z_2^2 z_3^2 = 1, \quad H_3(1,1,1) = \sum z_1^2 z_2$$

$$= \sum z_1^2 \sum z_1 - \sum z_1^3 = -3.$$

Thus

$$\det Q = c_0^3 H_3(3,0,0) + c_1^3 H_3(0,3,0) + c_2^3 H_3(0,0,3)$$
$$+ c_0 c_1 c_2 H_3(1,1,1)$$
$$= c_0^3 + c_1^3 + c_2^3 - 3c_0 c_1 c_2.$$

For the skew-symmetric case, $z_1' z_2' z_3' = -1$, $z'^3_j = -1$ and hence

$$\det Q' = c_0^3 - c_1^3 + c_2^3 + 3c_0 c_1 c_2.$$

§3   Case of composite n

It is pointed out by Muir ([4], p.472) that when $n = rs$, $\det Q$ can be expressed as the determinant of a circulant matrix of order r (or s). We give an explicit formula for this case.

Theorem 2. *Let $n = rs$, where r, s are positive integers. Let*

$$\beta = e^{2\pi i/r} = \omega^s, \tag{11}$$

$$\Psi(z) = p(z)p(\beta z) \cdots p(\beta^{r-1} z). \tag{12}$$

Then

$$\Psi(z) = F(z^r), \tag{13}$$

where F is a polynomial of degree at most $n - 1$. Let $q(z)$ be obtained from $F(z)$ by replacing each term bz^m by $bz^{m'}$, where m' is the residue of m (mod s). Then $q(z)$ has degree at most $s - 1$ and

$$\det Q = \det Q_s, \tag{14}$$

where Q_s is the circulant matrix of order s whose associated polynomial is $q(z)$.

Proof. By (2) we see that

$$\det Q = \Psi(z_1)\Psi(z_2) \cdots \Psi(z_s). \tag{15}$$

As above, we see that Ψ is a polynomial in z^r as in (13). Hence by (15)

$$\det Q = F(z_1^r) \cdots F(z_s^r). \tag{16}$$

Here F is being evaluated at the s roots of $z^s = 1$:

$$\zeta_1 = 1, \ \zeta_2 = \omega^r, \ldots, \ \zeta_s = \omega^{r(s-1)}. \tag{17}$$

Since $bz^m = bz^{m'}$ for $z = \zeta_j$ and $m \equiv m' \pmod{s}$, q and F agree at the ζ_j and

$$\det Q = q(\zeta_1)q(\zeta_2)\cdots q(\zeta_s) = \det Q_s, \tag{18}$$

as asserted. $\square$

Explicit formula for $q(z)$. We write

$$\gamma_1 = 1, \quad \gamma_2 = \beta, \ldots, \quad \gamma_r = \beta^{r-1}, \tag{19}$$

so that the γ_j are the roots of $z^r = 1$ and

$$\Psi(z) = p(\gamma_1 z)p(\gamma_2 z)\cdots p(\gamma_r z) = \sum_{m=0}^{n-1} h_m z^{rm}, \tag{20}$$

where

$$h_m = \sum c_0^{\alpha_0} c_1^{\alpha_1} \cdots c_{n-1}^{\alpha_{n-1}} \gamma_1^{l_1} \gamma_2^{l_2} \cdots \gamma_r^{l_r} \tag{21}$$

and the sum is over all (α) for which all $\alpha_j \geq 0$, $\sum_j \alpha_j = r$, $\sum_j j\alpha_j = mr$ and over all choices of $l_1, l_2, \ldots, l_r$ with α_0 of the l_i equal to 0, α_1 equal to $1, \ldots$. Thus

$$h_m = \sum_{(\alpha)} c_0^{\alpha_0} c_1^{\alpha_1} \cdots c_{n-1}^{\alpha_{n-1}} \ \ H_r(\alpha_0, \ldots, \alpha_{n-1}), \tag{22}$$

with the α_j restricted as above. Accordingly, $F(z) = \sum_m h_m z^m$ and

$$\begin{aligned}
q(z) = h_0 + h_s + \cdots + h_{(r-1)s} + z(h_1 + h_{s+1} + \cdots \\
+ h_{(r-1)s+1}) + \cdots + z^{s-1}(h_{s-1} + h_{2s-1} + \cdots + h_{rs-1}).
\end{aligned} \tag{23}$$

Example 2. $n = 6 = 2 \cdot 3$, $r = 2$, $s = 3$, $\beta = -1$, $\gamma_1 = 1$, $\gamma_2 = -1$, $h_0 = c_0^r = c_0^2$,

$$h_1 = \sum_{(\alpha)} c_0^{\alpha_0} \cdots c_5^{\alpha_5} \ \ H_2(\alpha_0, \ldots, \alpha_5),$$

with $\sum \alpha_j = 2$, $\sum j\alpha_j = 2$, $\alpha_j \geq 0$. The only cases for (α) are $(0, 2, 0, 0, 0, 0)$ and $(1, 0, 1, 0, 0, 0)$, so that

$$\begin{aligned}
h_1 &= c_1^2 H_2(0, 2, 0, 0, 0, 0) + c_0 c_2 H_2(1, 0, 1, 0, 0, 0) \\
&= -c_1^2 + 2c_0 c_2,
\end{aligned}$$

since $H_2(0,2,0,0,0,0) = G_2(1,1) = \sum \gamma_1 \gamma_2 = -1$, $H_2(1,0,1,0,0,0) = G_2(2,0) = \sum \gamma_1^2 = 2$. For h_2 we have $\sum j\alpha_j = 4$ and we obtain the (α) choices $(0, 0, 2, 0, 0, 0)$, $(1, 0, 0, 0, 1, 0)$, $(0, 1, 0, 1, 0, 0)$ and find

$$h_2 = c_2^2 + 2c_0 c_4 - 2c_1 c_3.$$

Similarly,

$$h_3 = -c_3^2 + 2c_2 c_4 - 2c_1 c_5, \quad h_4 = c_4^2 - 2c_3 c_5, \quad h_5 = -c_5^2.$$

Thus

$$q(z) = C_0 + C_1 z + C_2 z^2, \quad C_0 = h_0 + h_3, \quad C_1 = h_1 + h_4, \quad C_2 = h_2 + h_5.$$

Then by Ex. 1 above,

$$\det Q = C_0^3 + C_1^3 + C_2^3 - 3C_0 C_1 C_2.$$

Remark. In the skew-circulant case, (15), (16), (17) are modified by replacing Q by Q', z_j by z_j', ζ_j by $\zeta_j' = z'^r_j$, the roots of $z^s = -1$. We obtain $q_{sk}(z)$ by using the relationship $z^s = -1$ at these roots; thus bz^{m+ks} (with $m < s$) is replaced by $b(-1)^k z^m$ and

$$\det Q' = q_{sk}(\zeta_1') \cdots q_{sk}(\zeta_s') = \det Q_s', \tag{14}$$

where Q_s' is the skew-circulant matrix of order s whose associated polynomial is $q_{sk}(z)$. The explicit formula is as above, except that (23) becomes

$$q_{sk}(z) = h_0 - h_s + \cdots \pm h_{(r-1)s} + z(h_1 - h_{s+1} + \cdots) + \cdots. \tag{23'}$$

Example 3. For $n = 6 = 2 \cdot 3$, we use the results of Exs. 1 and 2 and obtain $q_{sk}(z) = C_0' + C_1' z + C_2' z^2$, where $C_0' = h_0 - h_3$, $C_1' = h_1 - h_4$, $C_2' = h_2 - h_5$, where the h_i are as in Ex. 2. Then, as at the end of Ex. 1,

$$\det Q' = \det Q_s' = C_0'^3 - C_1'^3 + C_2'^3 + 3C_0' C_1' C_2'.$$

§4 Functional analysis aspects

We can introduce several classes of related mappings, operating on a function $f(z)$:

$$\Psi_n(f) = G, \quad G(z) = f(z)f(\omega_n z) \cdots f(\omega_n^{n-1} z), \quad \omega_n = e^{2\pi i/n}; \quad (24)$$

$$\Phi_n(f) = F, \quad \text{where} \quad G(z) = F(z^n); \quad (25)$$

$$\mathcal{C}_n(f) = G(1) = F(1). \quad (26)$$

Here f can vary, for example, over the class of all polynomials, or over the class of all entire functions, or (for Ψ and Φ) over the class of all formal power series $\sum c_n z^n$. From the power series $\sum c_n z^n$ for f, one obtains a power series for $G = \Psi_n(f)$, containing only terms in z^{kn}; replacement of z^{kn} by z^k in all cases yields the series for F. The function G is a symmetrization of f, satisfying the identity $G(z) = G(\omega_n z)$.

The mappings are all multiplicative:

$$\left. \begin{aligned} \Psi_n(fg) = \Psi_n(f)\Psi_n(g), \quad \Phi_n(fg) = \Phi_n(f)\Phi_n(g), \\ \mathcal{C}_n(fg) = \mathcal{C}_n(f)\mathcal{C}_n(g). \end{aligned} \right\} \quad (27)$$

For polynomials the mappings can hence be analyzed by referring to the case of a constant function f_0 and a linear function $f_a = a + z$:

$$\Psi_n(f_0) = f_0^n = \Phi_n(f_0) = \mathcal{C}_n(f_0); \quad (28)$$

$$\left. \begin{aligned} \Psi_n(f_a) = a^n + (-1)^{n+1} z^n, \quad \Phi_n(f_a) = a^n + (-1)^{n+1} z, \\ \mathcal{C}_n(f_a) = a^n + (-1)^{n+1}. \end{aligned} \right\} \quad (29)$$

For entire functions one would be interested in the case of e^z or, more generally, of $\exp(az^m)$. For $f(z) = \exp(az^m)$ one finds

$$\Psi_n(f) = G, \; G(z) = \begin{cases} 1, & m \not\equiv 0 \pmod{n}, \\ \exp(naz^m), & m \equiv 0 \pmod{n}. \end{cases} \quad (30)$$

The main idea behind Theorem 2 is seen to be the identity

$$\Phi_{rs} = \Phi_r \circ \Phi_s, \quad (31)$$

which can be verified directly. Hence the operators Φ_1 (the identity), $\Phi_2, \Phi_3,$ ... form a multiplicative semigroup, isomorphic to the multiplicative semigroup of the positive integers.

The analogous identity for Ψ:

$$\Psi_{rs} = \Psi_r \circ \Psi_s \tag{32}$$

is valid only for r, s relatively prime, as one sees by elementary number theory.

§5 Another evaluation of $\det Q$

We can apply the ideas of the previous section to obtain another way of evaluating $\det Q$. Let the roots of $p(z) = c_0 + c_1 z + \cdots + c_{n-1}z^{n-1}$ be $-a_1, \ldots, -a_{n-1}$, so that

$$p(z) = c_{n-1}(z + a_1)\cdots(z + a_{n-1}). \tag{33}$$

Then by (2)

$$\det Q = \mathcal{C}(p). \tag{34}$$

By (33), (34), (28), (29), (31), we conclude that

$$\det Q = c_{n-1}^n(a_1^n + (-1)^{n+1})\cdots(a_{n-1}^n + (-1)^{n+1}). \tag{35}$$

The right side is a symmetric polynomial in $a_1, \ldots, a_{n-1}$. Hence it can be expressed as a polynomial in the elementary symmetric functions $E_1 = \sum a_1$, $E_2 = \sum a_1 a_2, \ldots, E_{n-1} = a_1 a_2 \cdots a_{n-1}$. But from (33)

$$E_1 = c_{n-2}/c_{n-1}, \quad E_2 = c_{n-3}/c_{n-1}, \ldots, \quad E_{n-1} = c_0/c_{n-1}. \tag{36}$$

Thus $\det Q$ is expressed as a function of $c_0, c_1, \ldots, c_{n-1}$ as desired.

To carry this out in detail, one has various alternatives. For example, from (35),

$$\det Q = c_{n-1}^n\left((-1)^{n+1} + \sum a_1^n + (-1)^{n+1}\sum a_1^n a_2^n \right.$$
$$\left. + \cdots + a_1^n a_2^n \cdots a_{n-1}^n\right). \tag{37}$$

The terms on the right side, after the first one, can be considered as elementary symmetric functions of $\beta_1 = a_1^n, \ldots, \beta_{n-1} = a_{n-1}^n$. We use a known identity ([2], p.14):

$$\sum \beta_1 \ldots \beta_l = \sum_{(\lambda)} \frac{(-1)^{\lambda_1 + \cdots + \lambda_l}}{\lambda_1! \cdots \lambda_l!} \left(\frac{\sigma_1}{1}\right)^{\lambda_1} \left(\frac{\sigma_2}{2}\right)^{\lambda_2} \cdots \left(\frac{\sigma_l}{l}\right)^{\lambda_l}, \tag{38}$$

where the sum is over all $(\lambda_1, \ldots, \lambda_l)$ with all $\lambda_j \geq 0$, $\lambda_1 + 2\lambda_2 + \cdots + l\lambda_l = l$ and $\sigma_p = \sum \beta_1^p$. Hence the right side of (37) can be expressed in terms of the power sums $s_{np} = \sum a_1^{np}$. Then Waring's formula expresses the power sums in terms of $E_1, \ldots, E_{n-1}$.

Example 4. $n = 3$. We have

$$(a_1^3 + 1)(a_2^3 + 1) = 1 + a_1^3 + a_2^3 + a_1^3 a_2^3$$
$$= 1 + (a_1 + a_2)^3 - 3a_1 a_2(a_1 + a_2) + a_1^3 a_2^3$$
$$= 1 + E_1^3 - 3E_1 E_2 + E_2^3.$$

Hence by (35) and (36)

$$\det Q = c_2^3 \left[1 + (c_1^3/c_2^3) - 3(c_1 c_0/c_2^2) + (c_0^3/c_2^3) \right]$$
$$= c_2^3 + c_1^3 - 3c_0 c_1 c_2 + c_0^3.$$

Here we could proceed without (38) and Waring's formula.

Remark. The material in ([4], p.471) appears to be related to this section.

§6 A result on symmetric functions

By equating two ways of evaluating $\det Q$, we obtain a theorem on symmetric functions:

Theorem 3. *For* $1 \leq k \leq n - 1$,

$$\prod_{j=1}^{k} \left[a_j^n + (-1)^{n+1} \right] = \sum_{(\alpha)} E_1^{\alpha_{k-1}} E_2^{\alpha_{k-2}} \cdots E_k^{\alpha_0} H_n(\alpha_0, \ldots, \alpha_k, 0, \ldots, 0) \quad (39)$$

where $(\alpha) = (\alpha_0, \ldots, \alpha_k, 0, \ldots, 0)$ *satisfies* (8) *and* $E_1 = \sum a_1$, $E_2 = \sum a_1 a_2$, *... are the elementary symmetric functions of* $a_1, \ldots, a_k$.

Proof. We take $p(z) = (z + a_1) \cdots (z + a_k) = c_0 + c_1 z + \cdots$, so that $c_0 = E_k$, $c_1 = E_{k-1}, \ldots, c_{k-1} = E_1$, $c_k = 1$ and $c_j = 0$ for $k < j \leq n - 1$. The left side of (39) is then $\det Q$, as in Section 5; the right side is obtained from (7). $\square$

§7 Extension to the skew-symmetric case

The discussion of Sections 4–6 extends easily to the skew-symmetric case. In (24) one defines $G(z)$ as

$$f(e^{\pi i/n}z)f(e^{\pi i/n}\omega_n z)\cdots f(e^{\pi i/n}\omega_n^{n-1}z)$$

and (27), (28) continue to hold. In (29) and (35) $(-1)^{n+1}$ is replaced by $(-1)^n$; in (30) $\exp(naz^m)$ is replaced by $\exp(-naz^m)$. In (39) $(-1)^{n+1}$ is replaced by $(-1)^n$ and H_n by H_n'. The identity (31) becomes

$$\Phi'_{rs} = \Phi'_r \circ \Phi_s, \tag{31'}$$

where Φ'_n is the modification of Φ_n obtained from the new $G(z)$.

References

[1] Philip J. Davis, Circulant Matrices, John Wiley & Sons, New York, 1979.

[2] Francesco Faà de Bruno, Théorie des Formes Binaires, Librairie Brero, Turin, 1876.

[3] Oystein Ore, Some studies on cyclic determinants, Duke Math. Journal vol. 51 (1951), pp. 343-354.

[4] Thomas Muir, A Treatise on the Theory of Determinants, Longmans, Green & Co., New York, 1933.

Department of Mathematics
University of Michigan
Ann Arbor, MI 48109, U.S.A.

J. Korevaar

Interpolation by Entire Functions in $\mathbb{C}$ — another Look

§1 Introduction

In the early part of his mathematical career, Albert Pfluger wrote a number of papers involving interpolation by entire functions of restricted growth [8, 9, 10]. Here interpolation was studied both as a tool and for its own sake. Constructing explicit interpolation formulas, Pfluger and others, notably N. Levinson, obtained very precise results; for details, see the books by R.P. Boas [2] and B. Ya. Levin [6].

A standard question in contemporary complex analysis asks to extend classical results to the case of $\mathbb{C}^n$, where explicit interpolation formulas can rarely be obtained. There one usually has to be satisfied with existence proofs for interpolating functions of appropriately limited growth. Such proofs may be derived from L. Hörmander's existence theory [5] for solutions of the $\bar{\partial}$-problem (inhomogeneous Cauchy-Riemann equations) in weighted L^2 spaces, see [3, 4, 11]. It would be worth looking for further applications involving interpolation.

It is not well known that Hörmander's theory can also be very effective in the case of $\mathbb{C}$. We mention B. Berndtsson's paper [1]; as another illustration, we will now discuss a problem related to one treated by Pfluger [9]. The general question goes back to G. Pólya and V. Bernstein: Considering analytic functions of exponential type in an angle around the positive real axis $\mathbb{R}^+$ and sequences $\lambda = \{\lambda_n\}$ on or near $\mathbb{R}^+$, one asks for conditions under which the growth of the functions on λ determines their growth on $\mathbb{R}^+$.

We will use Hörmander's theory to give a new proof of

Theorem 1. *Let f be analytic and of exponential type on the closed right half-plane $H : \{\operatorname{Re} z \geq 0\}$ and bounded on the imaginary axis $\mathbb{I}$. Let $\lambda = \{\lambda_n\}$,*

$n = 1, 2, \ldots$ *be a strictly increasing sequence of positive real numbers with* $\sum 1/\lambda_n = \infty$ *that satisfies the uniform separation condition*

$$\sigma_n \overset{\text{def}}{=} -\log \prod_{\lambda_n/2 < \lambda_k < 2\lambda_n,\, k \neq n} |1 - \lambda_n/\lambda_k| = \mathrm{o}(\lambda_n), \quad n \to \infty. \tag{1.1}$$

Then the type of f *on* $\mathbb{R}^+$ *equals its type on* λ*:*

$$\limsup_{x \to +\infty} \frac{1}{x} \log |f(x)| = \limsup_{n \to \infty} \frac{1}{\lambda_n} \log |f(\lambda_n)|.$$

This refined result is essentially due to Levinson, see [7] and cf. Boas [2]. The boundedness of f on $\boldsymbol{I}$ can be relaxed to integrability of $(1+z)^{-2} \log^+ |f|$ without much effort, cf. the footnote on p. 118 of [7]. The present modern separation condition allows a little more clustering than Levinson's condition; both are satisfied if λ consists of positive integers and has density zero. One can prove a related result for the case where λ is a sequence of distinct points with $|\lambda_n| \to \infty$ and $\arg \lambda_n \to \alpha$, $|\alpha| < \pi/2$, cf. R.L. Zeinstra [12]. That Ph.D. thesis also discusses the connection with the regular growth results of L. Ahlfors – M. Heins and V.S. Azarin for bounded subharmonic functions.

§2 Solution of $\overline{\partial}$ and interpolation with growth conditions

We only consider the case of one complex variable, for which Hörmander's theorem is relatively easy.

Theorem 2 (cf. Hörmander [5], Theorem 4.4.2). *Let* β *be any subharmonic function on* $\mathbb{C}$ *and let* v *be a function of class* C^1 *such that* $\int_{\mathbb{C}} |v|^2 e^{-\beta} dm$ *is finite, where* m *denotes planar Lebesgue measure. Then there is a* C^1 *solution* u *of the inhomogeneous Cauchy-Riemann equation* $\partial u / \partial \bar{z} = v$ *on* $\mathbb{C}$ *which satisfies the growth condition*

$$2 \int_{\mathbb{C}} |u|^2 e^{-\beta} (1 + |z|^2)^{-2} dm \leq \int_{\mathbb{C}} |v|^2 e^{-\beta} dm. \tag{2.1}$$

Suppose now that $\lambda = \{\lambda_n\}$, $n = 1, 2, \ldots$ is a sequence of distinct points tending to infinity in $\mathbb{C}$ and that $b = \{b_n\}$ is an arbitrary sequence of complex numbers. In order to obtain a holomorphic function h on $\mathbb{C}$ such that

$$h(\lambda_n) = b_n, \quad n = 1, 2, \ldots,$$

one may first determine a C^2 solution g of the interpolation problem. To that end, let ρ be an arbitrary C^2 function on $\mathbb{C}$ such that

$$\rho(z) = \begin{cases} 1 & \text{for } |z| \leq 1/4, \\ 0 & \text{for } |z| \geq 1/2. \end{cases}$$

Choosing numbers d_n such that

$$0 < d_n \leq \min_k |\lambda_n - \lambda_k|, \tag{2.2}$$

so that the discs $B(\lambda_n, d_n/2)$ are disjoint, one may take

$$g(z) \overset{\text{def}}{=} \sum_1^\infty b_n \rho((z - \lambda_n)/d_n). \tag{2.3}$$

One next seeks to modify the interpolating C^2 function g so as to obtain a holomorphic solution

$$h = g - u \tag{2.4}$$

of the interpolation problem. Analyticity of h is equivalent to the requirements $h \in C^1$ and $\partial h/\partial \bar{z} = 0$. Thus u will be an appropriate correction term if and only if

$$u \in C^1, \quad \partial u/\partial \bar{z} = \partial g/\partial \bar{z} \quad \text{and} \quad u(\lambda_n) = 0, \quad n = 1, 2, \ldots. \tag{2.5}$$

The existence of solutions u follows from Theorem 2: there are subharmonic functions β of arbitrarily rapid growth, hence the last integral in (2.1), with $v = \partial g/\partial \bar{z}$, can be made finite. To assure $u = 0$ on the sequence λ it suffices, as was observed by E. Bombieri [3], to give $e^{-\beta}$ a nonintegrable singularity at each point λ_n. One may for example require that

$$\beta(z) \leq 2\log|1 - z/\lambda_n| + \mathcal{O}(1) \quad \text{for} \quad z \to \lambda_n. \tag{2.6}$$

One naturally wants to obtain a solution u and corresponding function h of (approximately) minimal growth. In view of (2.1), one thus wants β to have small growth. Here (2.6) and the growth of $v = \partial g/\partial \bar{z}$ play a role; the latter depends on the interplay of λ and b.

§3   An auxiliary interpolating function for Theorem 1

Let f, H, I and λ be as in Theorem 1. Multiplying f by a suitable real exponential function, we may assume that it is of negative type along $I\!R^+$. Since f is of exponential type on H and bounded on I, it follows that f will be bounded on H. We now choose $a > 0$ such that

$$\text{type}_\lambda f < -a. \tag{3.1}$$

We will use the considerations of Section 2 to prove that there is an interpolating entire function h,

$$h(\lambda_n) = b_n \overset{\text{def}}{=} e^{a\lambda_n} f(\lambda_n), \quad n = 1, 2, \ldots \tag{3.2}$$

which does not grow much faster than

$$p(z) \overset{\text{def}}{=} \prod (1 - z^2/\lambda_n^2). \tag{3.3}$$

The function $v = \partial g/\partial \bar{z}$ may be estimated from (2.3). The nth term $b_n(\partial/\partial \bar{z})\rho((z - \lambda_n)/d_n)$ has its support in the annulus A_n: $d_n/4 \leq |z - \lambda_n| \leq d_n/2$ and the different annuli do not overlap. Since $\partial \rho/\partial \bar{z}$ is bounded,

$$|v(z)| = |\partial g/\partial \bar{z}| \leq C|b_n|/d_n, \quad z \in A_n. \tag{3.4}$$

Note that by (3.1), (3.2) there will be a number $\delta > 0$ such that

$$|b_n| \leq e^{-\delta\lambda_n}, \quad n > n_0. \tag{3.5}$$

The separation condition (1.1) shows that for $\lambda_n/2 < \lambda_k < 2\lambda_n$ one has $\log|\lambda_k - \lambda_n| > - \text{o}\,(\lambda_n)$, hence for any given $\epsilon > 0$ we may take d_n in (2.2) equal to $\exp(-\epsilon\lambda_n)$, $n > n_0$. The condition also implies that the number of points λ_k on $(\lambda_n/2, 2\lambda_n)$ equals $\text{o}(\lambda_n)$, so that $n/\lambda_n \to 0$ and $p(z)$ will be entire and of exponential type zero.

Considering (2.6) we choose the subharmonic function β equal to $2\log|p|$; the additional singularities at the points $-\lambda_n$ help limit the growth of β. We proceed to estimate $e^{-\beta}$ on the annuli A_n. By the preceding, for $z \in A_n$ and $n > n_0$,

$$\prod_{\lambda_k < 2\lambda_n} |1 - z^2/\lambda_k^2| \geq \prod_{\lambda_n/2 < \lambda_k < 2\lambda_n} \geq d_n \prod_{\lambda_n/2 < \lambda_k < 2\lambda_n}^{(k \neq n)} |1 - \lambda_n/\lambda_k|,$$

$$\log \prod_{\lambda_k \geq 2\lambda_n} |1 - z^2/\lambda_k^2| = \int_{2\lambda_n-}^{\infty} \log|1 - z^2/t^2| dn(t)$$

$$\geq -C \int_{2\lambda_n-}^{\infty} (\lambda_n^2/t^2) dn(t) = o(\lambda_n).$$

Thus in view of the separation condition (1.1), for any $\epsilon > 0$

$$e^{-\beta} = |p(z)|^{-2} \leq C e^{\epsilon \lambda_n}, \quad z \in A_n, \quad n = 1, 2, \ldots . \tag{3.6}$$

Combining (3.4)–(3.6) with $\epsilon = \delta$, we find that for our v and β, the last integral in (2.1) is finite:

$$\int_{\mathbb{C}} |v|^2 e^{-\beta} dm = \sum_1^{\infty} \int_{A_n} \leq \sum_1^{\infty} C|b_n|^2 d_n^{-2} e^{\delta \lambda_n} d_n^2 \leq C \sum_1^{\infty} e^{-\delta \lambda_n}.$$

Now let u be a solution of the corresponding problem (2.5) as given by Theorem 2 and let $h = g - u$ (2.4), so that $h = -u$ outside the small discs $B(\lambda_n, d_n/2)$. The finiteness of the first integral (2.1) and the mean-value theorem for h on discs, combined with Schwarz's inequality, then give an estimate for $|h(z)|$ away from the sequence λ. For z outside an angular neighborhood of $\mathbb{R}^+$ one readily finds that

$$|h(z)| \leq C e^{\beta/2}(1 + |z|^2) = C|p(z)|(1 + |z|^2). \tag{3.7}$$

On the whole plane, h will be of exponential type zero. (An explicit interpolation formula would give somewhat sharper inequalities.)

§4 Use of quasi-analyticity to complete the proof of Theorem 1

The following argument is a simplification of Levinson's [7]. We continue the discussion of Section 3. By (3.7) the quotient h/p is analytic and polynomially bounded on some strip around the imaginary axis $\mathbf{I}$. Thus h/p is the Fourier-Laplace transform of a tempered distribution on $\mathbb{R}$, but distributions can easily be avoided. To that end we may replace h/p by h/q, where

$$q(z) = (1 + z)^4 p(z) = (1 + z)^4 \prod(1 - z^2/\lambda_n^2). \tag{4.1}$$

By (3.7), $h/q = \mathcal{O}((1 + z)^{-2})$ away from $\mathbb{R}^+$, hence

$$\varphi(t) \overset{\text{def}}{=} \frac{1}{2\pi i} \int_{\mathbf{I}} \frac{h(z)}{q(z)} e^{tz} dz, \quad t \in \mathbb{R} \tag{4.2}$$

is a bounded continuous function with Laplace transform

$$\mathcal{L}\varphi(z) = \int_{I\!R} \varphi(t)e^{-zt}dt$$

equal to h/q on a strip around $I\!\!I$. Slightly bending the upper and lower half of the path of integration in (4.2) to the right, one sees that $\varphi(t)$ is analytic for $t < 0$. (It will vanish for $t > 0$.)

The function h agrees with $e^{az}f$ on the sequence λ (3.2); we want to compare f/q with $e^{-az}h/q$. The latter function will be the Laplace transform of $\varphi(t - a)$ and by the above, $\varphi(t - a)$ is analytic for $t < a$.

We now consider f/q. Since f is analytic and bounded on the half-plane H, f/q is analytic and small at ∞ on some strip $0 \le \mathrm{Re}\, z < c$. Thus f/q is the Laplace transform $\mathcal{L}\Psi$ of a C^∞ function:

$$\Psi(t) \stackrel{\text{def}}{=} \frac{1}{2\pi i} \int_{I\!\!I} \frac{f(z)}{q(z)} e^{tz} dz, \quad t \in I\!R. \tag{4.3}$$

We will show that Ψ belongs to a quasi-analytic (q.a.) class, specifically, $\Psi \in C(M_n)$ where $M_{2k} = \lambda_1^2 \dots \lambda_k^2$, $M_{2k+1} = M_{2k}\lambda_{k+1}$. Indeed, by (4.3) and (4.1), for all $t \in I\!R$ and all $k \ge 0$:

$$|\Psi^{(2k)}(t)| \le \frac{1}{2\pi} \int_{I\!R} \frac{|f(iy)|y^{2k}\,dy}{(1+y^2)^2 \prod_{n=1}^{k}(1+y^2/\lambda_n^2)} \le C\lambda_1^2 \dots \lambda_k^2,$$

and similarly for $\Psi^{(2k+1)}(t)$. That $C(M_n)$ is a q.a. class follows immediately from the Denjoy-Carleman theorem: the series $\sum M_n/M_{n+1}$ is divergent since by the hypothesis of Theorem 1, the series $\sum 1/\lambda_k$ is divergent.

We finally compare f/q and $e^{-az}h/q$:

$$(f - e^{-az}h)/q = \mathcal{L}\omega, \quad \omega(t) = \Psi(t) - \varphi(t - a). \tag{4.4}$$

By (3.2) the function on the left is analytic on H. Since it is of exponential type and $\mathcal{O}((1 + z)^{-2})$ for $\pi/4 \le |\arg z| \le \pi/2$, it will be $\mathcal{O}((1 + z)^{-2})$ throughout H. It follows that $\omega(t) = 0$ for $t < 0$. We know that Ψ belongs to the q.a. class $C(M_n)$ on $I\!R$. Now by the analyticity of $\varphi(t - a)$ for $t < a$, that function will belong to $C(M_n)$ on every compact subset of $(-\infty, a)$ (since $n/\lambda_n \to 0$). Thus also $\omega \in C(M_n)$ on those compact subsets and hence $\omega(t) = 0$ on $(-\infty, a)$.

By (4.4), it now follows that

$$f(x) = q(x) \int_a^\infty \omega(t)e^{-xt}dt + e^{-ax}h(x), \quad x > 0. \tag{4.5}$$

Here q and h are of exponential type 0 and ω is bounded. The final conclusion is that f is of type $\leq -a$ on $\mathbb{R}^+$. In combination with (3.1), this result completes the proof of Theorem 1.

References

[1] B. Berndtsson, A note on Pavlov-Korevaar-Dixon interpolation. Nederl. Akad. Wetensch. Proc. Ser. A 81 (1978) 409–414.

[2] R.P. Boas, Entire functions. Academic Press, New York, 1954.

[3] E. Bombieri, Algebraic values of meromorphic functions. Inventiones Math. 10 (1970) 267–287.

[4] L. Gruman, Interpolation in spaces of entire functions in $\mathbb{C}^N$. Canad. Math. Bull. 19 (1976) 109–112.

[5] L. Hörmander, An introduction to complex analysis in several variables. North-Holland, Amsterdam. 1973.

[6] B.Ya. Levin, Distribution of zeros of entire functions. Amer. Math. Soc. Transl. of Math. Mono. 5, Providence, 1964.

[7] N. Levinson, Gap and density theorems. Amer. Math. Soc. Colloq. Publ. 26, New York, 1940.

[8] A. Pfluger, On analytic functions bounded at the lattice points. Proc. London Math. Soc. (2) 42 (1936) 305–315.

[9] A. Pfluger, Über das Anwachsen von Funktionen, die in einem Winkelraum regulär und vom Exponentialtypus sind. Compositio Math. 4 (1937) 367–372.

[10] A. Pfluger, Über Interpolation ganzer Funktionen. Comment. Math. Helv. 14 (1942) 314–349.

[11] H. Skoda, Sous-ensembles analytiques d'ordre fini ou infini dans $\mathbb{C}^n$. Bull. Soc. Math. France 100 (1972) 353–408.

[12] R.L. Zeinstra, Müntz-Szász approximation on curves and area problems for zero sets. Ph. D. thesis, Univ. of Amsterdam, 1985.

Universiteit van Amsterdam
Math. Institute
Roetersstraat 15
1018 WB Amsterdam
The Netherlands

Reiner Kühnau

Möglichst konforme Spiegelung an einem Jordanbogen auf der Zahlenkugel

§1 Einleitung

Ein abgeschlossener Jordanbogen $\mathfrak{r}$ mit Endpunkten z_1 and z_2 auf der Riemannschen z-Zahlenkugel heiße "quasikonformer Spiegel", wenn eine quasikonforme Spiegelung an $\mathfrak{r}$ existiert, d.h. eine orientierungsumkehrende quasikonforme Abbildung der Zahlenkugel auf sich, bei der $\mathfrak{r}$ punktweise festbleibt. (Dabei wird nicht verlangt, daß der unendlich ferne Punkt festbleibt.) Eine quasikonforme Spiegelung an $\mathfrak{r}$ mit kleinstmöglicher Dilatationsschranke heißt "möglichst konform" im Anschluß an die Sprechweise von H. Grötzsch oder "extremal quasikonform" im Anschluß an O. Teichmüller. Die zugehörige kleinstmögliche Dilatationsschranke $Q_{\mathfrak{r}} \geq 1$ bzw. $q_{\mathfrak{r}} = (Q_{\mathfrak{r}} - 1)/(Q_{\mathfrak{r}} + 1)$ nennen wir "Spiegelungskoeffizienten" von $\mathfrak{r}$. Dieser ist invariant bei linearer Transformation.

Ganz entsprechende Begriffsbildungen, insbesondere der Spiegelungskoeffizient $Q_{\mathfrak{C}}$ bzw. $q_{\mathfrak{C}}$ werden für geschlossene Jordankurven $\mathfrak{C}$ betrachtet. Man vgl. zu diesem auf L. V. Ahlfors zurückgehenden Fragenkreis [11], [3], [10].

Bildet man die zweiblättrige Riemannsche Fläche mit Windungspunkten bei z_1 und z_2 konform auf die schlichte Zahlenkugel ab, entsteht aus $\mathfrak{r}$ eine geschlossene Jordankurve $\mathfrak{C}$. Aus einer Q-quasikonformen Spiegelung an $\mathfrak{r}$ entsteht so eine Q-quasikonforme Spiegelung an $\mathfrak{C}$ und umgekehrt. Insbesondere gilt $Q_{\mathfrak{C}} = Q_{\mathfrak{r}}$. Es ist $q_{\mathfrak{C}} = 0$ durch elementare funktionentheoretische Schlußweise genau für eine Kreislinie $\mathfrak{C}$. Daraus fließt: Es ist $q_{\mathfrak{r}} = 0$ genau für einen Kreisbogen $\mathfrak{r}$.

Durch diese Zurückführung auf den Fall geschlossener Jordankurven $\mathfrak{C}$ ergibt sich weiter (vgl. z.B. [10]): Besteht $\mathfrak{r}$ in Umgebung von $z_0 \in \mathfrak{r}$ aus zwei dort unter dem Winkel $\alpha\pi$ zusammenstoßenden abgeschlossenen analytischen

Bögen, dann gilt die "Knickbedingung"

$$q_{\mathfrak{r}} \geq |1 - \alpha|. \tag{1}$$

Dies liefert uns $q_{\mathfrak{r}} = |1 - \alpha|$ für den Fall, $\mathfrak{r}$ besteht aus zwei unter dem Winkel $\alpha\pi$ zusammenstoßenden Strecken. Denn hier kann man sofort eine Spiegelung mit der rechten Seite von (1) als Betrag der komplexen Dilatation angeben; diese ist sogar Spiegelung an dem aus den beiden Trägerstrahlen bestehenden $\mathfrak{C}$ (vgl. [10]).

Man kann noch andere $\mathfrak{r}$ mit Knick angeben, bei denen so $q_{\mathfrak{r}}$ über (1) gewinnbar ist. Jedoch gibt es anscheinend derzeit sonst — abgesehen von dem trivialen Beispiel des Kreisbogens — kein einziges $\mathfrak{r}$ mit explizit bekanntem $q_{\mathfrak{r}}$.

Durch diese Zurückführung auf den Fall geschlossener Jordankurven $\mathfrak{C}$ folgt ferner zu jedem $\mathfrak{r}$ die Existenz mindestens einer möglichst konformen Spiegelung. Daß diese für z.B. analytisches $\mathfrak{r}$ eindeutig bestimmt ist und sich durch ein quadratisches Differential charakterisieren läßt, folgt durch die Strebelsche Theorie möglichst konformer Abbildungen bei fest gegebenen Randwerten [16], [10], wie sich auch manches andere darauf zurückführen läßt.

Jedoch ergeben sich auch Fragen, die sich bei der Spiegelung an einem Jordanbogen in natürlicherer Weise stellen als bei möglichst konformen Abbildungen mit festen Randwerten. Dies sieht man z.B. an dem folgenden Satz, der unser Hauptziel darstellt.

Satz 1. *Es sei $\mathfrak{r}$ ein abgeschlossener Teilbogen eines festen analytischen Jordanbogens, wobei $\mathfrak{r}$ die Länge 2ε besitzt und diese Länge durch den festen Punkt $z_0 \in \mathfrak{r}$ halbiert wird. Dann gilt für $\varepsilon \to 0$*

$$q_{\mathfrak{r}} = \frac{1}{12}|dk/ds| \cdot \varepsilon^2 + O(\varepsilon^3), \tag{2}$$

wobei dk/ds die Ableitung der Krümmung k nach der Bogenlänge s im Punkte z_0 ist. $O(\varepsilon^3)/\varepsilon^3$ bleibt für $\varepsilon \to 0$ beschränkt.

Auch zur Geometrie der möglichst konformen Spiegelung an $\mathfrak{r}$ läßt sich für hinreichend kleine ε etwas aussagen, falls $dk/ds \neq 0$ in z_0.

Satz 2. *Für hinreichend kleine ε besitzt das die möglichst konforme Spiegelung an $\mathfrak{r}$ beschreibende quadratische Differential außerhalb und auf den beiden Ufern von $\mathfrak{r}$ keine Nullstelle.*

Hierdurch können wir wie folgt $q_{\mathfrak{r}}$ für kleine ε auch anders charakterisieren.

Als reziproken Fredholmschen Eigenwert $\kappa_{\mathfrak{r}}$ des Bogens $\mathfrak{r}$ definieren wir zunächst

$$\kappa_{\mathfrak{r}} = \sup_{h_1, h_2} \frac{|D(h_1) - D(h_2)|}{D(h_1) + D(h_2)}. \tag{3}$$

Dabei ist das Supremum zu nehmen über alle Paare von außerhalb $\mathfrak{r}$ (incl. in ∞) harmonischen Funktionen h_1, h_2, die noch bei Annäherung an $\mathfrak{r}$ stetige Randwerte besitzen (diese Randwerte können für die beiden Uferseiten von $\mathfrak{r}$ verschieden sein), wobei die Randwerte von h_1 auf den beiden Ufern von $\mathfrak{r}$ jeweils mit den Randwerten von h_2 auf den gegenüberstehenden Ufern von $\mathfrak{r}$ übereinstimmen müssen. Mit $D(..)$ wird das Dirichletsche Integral über die gesamte Ebene außerhalb $\mathfrak{r}$ bezeichnet. Diese Definition von $\kappa_{\mathfrak{r}}$ wird nahegelegt durch die entsprechende Definition bei geschlossenen Jordankurven (vgl. z.B. die Darstellung in [15]), auf welch letzteren Fall wir zurückführen können, indem wir die zweiblättrige Riemannsche Fläche mit Windungspunkten bei den Endpunkten von $\mathfrak{r}$ auf die schlichte Ebene schlicht konform abbilden. Der gemäß (3) definierte reziproke Fredholmsche Eigenwert von $\mathfrak{r}$ und der der dabei entstehenden geschlossenen Jordankurve $\mathfrak{C}$ stimmen wegen der Invarianz der Dirichletschen Integrale bei konformer Abbildung natürlich überein. Da Entsprechendes nach Obigem für $q_{\mathfrak{r}}$ und $q_{\mathfrak{C}}$ gilt, haben wir nach einer klassischen Ungleichung von L. V. Ahlfors (vgl. [1], [2], S.36, [10])

$$\kappa_{\mathfrak{r}} \leq q_{\mathfrak{r}}.$$

In unserem Falle gilt nun genauer bei $dk/ds \neq 0$ in z_0 der

Satz 3. *Für alle hinreichend kleinen ε ist*

$$\kappa_{\mathfrak{r}} = q_{\mathfrak{r}}. \tag{4}$$

Über Satz 2 noch hinausgehend gilt über die Geometrie der möglichst konformen Spiegelung an $\mathfrak{r}$ (wieder bei $dk/ds \neq 0$ in z_0) der folgende Satz. Zu dessen bequemeren Formulierung führen wir zunächst eine feste (d.h. von ε unabhängige) lineare Transformation mit $z_0 \to 0$ in eine $\mathfrak{z}$-Ebene so aus, daß der Krümmungskreis in z_0 in eine Gerade übergeht, und anschließend noch eine (von ε abhängende) ganz-lineare Transformation in eine Z-Ebene, bei der dann die Endpunkte des Bildes $\mathfrak{r}_Z$ von $\mathfrak{r}$ bei ± 2 landen. Dabei können wir (für hinreichend kleine ε) o.E.d.A. (sonst nähmen wir eine Spiegelung an der reellen Achse vor) annehmen, daß $\mathfrak{r}_Z$ im Bildpunkte von z_0 einen negativen Anstieg hat (vgl. Fig. 1).

z_1
z_2
ε
ε
r
z_0
z — Ebene
r_z
z_1
O
z — Ebene
z_2
r Z
−2
+2
Z — Ebene
−1
C_w
+1
w — Ebene
e^{iπ/4}
−e^{iπ/4}
C_w
w — Ebene
C_W
W — Ebene
|W| = 1
W — Ebene

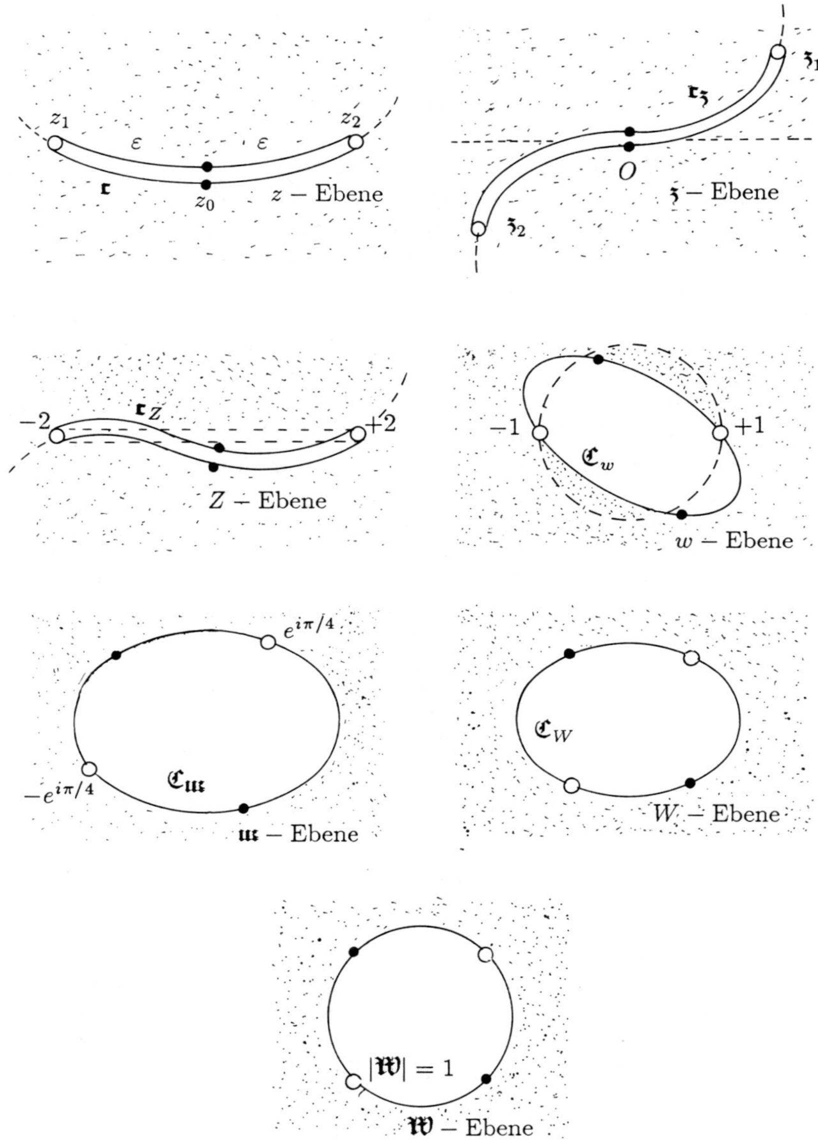

Figur 1.

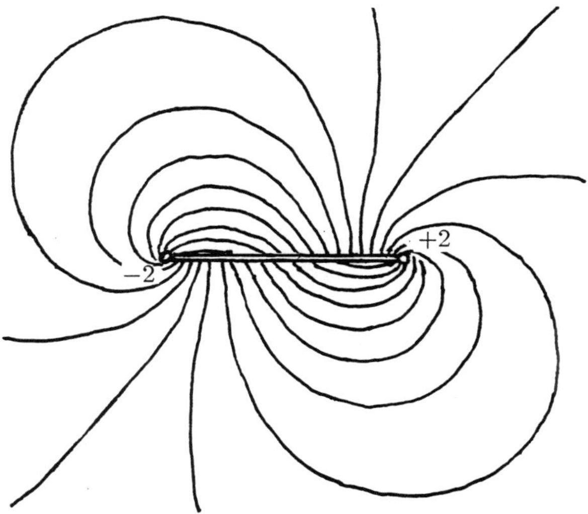

Figur 2. Schar S in der Z-Ebene im Grenzfall $\varepsilon=0$

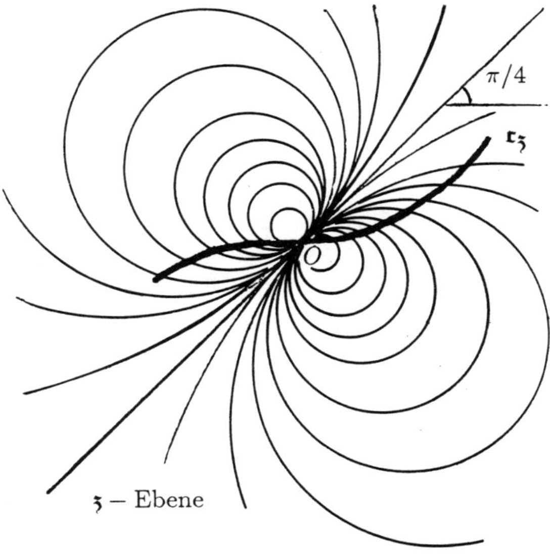

Figur 3.

Sodann wird das Äußere von $\mathfrak{r}_Z$ auf das Innere des Einheitskreises mit $\infty \to 0$ und mit der Entwicklung $a/Z + \ldots (a > 0)$ in $Z = \infty$ schlicht konform abgebildet. Als Schar S bezeichnen wir in der Z-Ebene die Gesamtheit der Urbilder (bei dieser schlichten konformen Abbildung) der Strecken des Neigungswinkels $+3\pi/4$ gegen die positiv-reelle Achse innerhalb des Einheitskreises.

Satz 4. *Bei der möglichst konformen Spiegelung an $\mathfrak{r}_Z$ gehen infinitesimale Kreise in infinitesimale Ellipsen des Achsenverhältnisses $Q_\mathfrak{r}$ über, wobei die großen Achsen einen Neigungswinkel besitzen, der in jedem Punkte außerhalb $\mathfrak{r}_Z$ übereinstimmt mit dem Neigungswinkel der durch diesen Punkt verlaufenden Kurve von S bis auf eine additive Fehlergröße, die nach Division durch ε beschränkt in Z (außerhalb $\mathfrak{r}_Z$) und ε (hinreichend klein) ist.*

Die Schar S ist in Figur 2 dargestellt im Grenzfalle $\varepsilon = 0$. Diese Schar entsteht dann also als Urbild der Strecken des Neigungswinkels $+3\pi/4$ gegen die positiv-reelle Achse (innerhalb des Einheitskreises) bei schlichter konformer Abbildung des Äußeren der Strecke $-2 \ldots +2$ auf das Innere des Einheitskreises mit $\infty \to 0$ und der Entwicklung $1/Z + \ldots$ in $Z = \infty$. Bei den in $\pm\sqrt{2}$ endenden Kurven von S handelt es sich übrigens um halbe Äste der zu ± 2 konfokalen gleichseitigen Hyperbeln. Die Schar S ist zu 0 zentrisch symmetrisch. Bei Spiegelung an der reellen Achse entsteht aus S die Schar der orthogonalen Trajektorien. Im oberen Ufer der Strecke $-2 < Z < +2$ endet die in Z einmündende Kurve von S dort mit dem Winkel $\pi/4 + \arccos(Z/2)$.

Eine zu Satz 4 entsprechende Aussage ergibt sich für die $\mathfrak{z}$-Ebene, wobei dann die Neigungen der großen Achsen der infinitesimalen Bildellipsen analog außerhalb eines beliebig kleinen aber festen zu $\mathfrak{z} = 0$ konzentrischen Kreises approximiert werden durch die Schar der Figur 3 (parabolisches Kreisbüschel der Neigung $\pi/4$).

Abschließend noch zwei allgemeine Bemerkungen.

1.) Neben der in dieser Mitteilung betrachteten Fragestellung der "quasikonformen Spiegel" steht die Frage nach "quasikonformen Jordanbögen" $\mathfrak{r}$, die Bild einer Strecke bei einer quasikonformen Abbildung der Vollebene sind [11] (S.101). Für die hierbei kleinstmögliche Dilatationsschranke $Q_\mathfrak{r}^*$ bzw. $q_\mathfrak{r}^* = (Q_\mathfrak{r}^* - 1)/(Q_\mathfrak{r}^* + 1)$ (etwa "Streckungskoeffizient" von $\mathfrak{r}$ zu nennen) gilt $Q_\mathfrak{r} \leq Q_\mathfrak{r}^{*2}$, da man aus einer Q-quasikonformen Abbildung der Vollebene, die $\mathfrak{r}$ in eine Strecke überführt (durch Anwendung hin- und rückwärts, dazwischen eine gewöhnliche Spiegelung), eine Q^2-quasikonforme Spiegelung an $\mathfrak{r}$ konstruieren kann. Eine genauere Klärung der Zusammenhänge (die offenbar komplizierter als bei geschlossenen Jordankurven sind), steht noch aus.

2.) Es wäre sicher interessant, ähnliche Zusammenhänge — wie in den Sätzen 1 bis 4 dargestellt — zwischen differentialgeometrischen Größen einerseits und andererseits dem infinitesimalen Verhalten der Dilatationsschranke der möglichst konformen Spiegelungen auch in höheren Dimensionen zu suchen. (Die hier in dieser Mitteilung verwendeten Methoden versagen dann wohl allerdings vollständig.) Eine interessante Frage in diesem Zusammenhange wäre übrigens noch (schon im Falle der Ebene): Wie kann man diejenigen Punktmengen charakterisieren, zu denen eine quasikonforme Spiegelung existiert, d.h. eine orientierungsumkehrende quasikonforme Abbildung der Vollebene, die die Punkte dieser Menge einzeln festhält?

§2 Einfache Vorbemerkungen

Es sei $\mathfrak{r}$ ein Jordanbogen oder eine geschlossene Jordankurve mit der Parameterdarstellung $z(t)$, $a \leq t \leq b$ ($z(a) = z(b)$, falls $\mathfrak{r}$ geschlossen ist). Wählen wir einen Teilbogen von $\mathfrak{r}$ gemäß $(a \leq)t_1^* \leq t \leq t_2^*(\leq b)$, besitzt dieser einen gewissen Spiegelungskoeffizienten $q(t_1^*, t_2^*)$.

Monotonieeigenschaft: Es gilt für $t_1^* \leq t_1 < t_2 \leq t_2^*$

$$q(t_1, t_2) \leq q(t_1^*, t_2^*), \tag{5}$$

insbesondere stets $q(t_1, t_2) \leq q_{\mathfrak{r}}$.

(Gilt die entsprechende Aussage auch für den κ-Wert von (2)?)

Denn jede Q-quasikonforme Spiegelung an einem Bogen stellt auch eine Q-quasikonforme Spiegelung an einem Teilbogen dar.

In (5) kann das Gleichheitszeichen stehen, obwohl $t_1^* \neq t_1$ und $t_2 \neq t_2^*$. Man vergleiche (neben dem trivialen Beispiel eines Kreisbogens) das in §1 genannte Beispiel, $\mathfrak{r}$ besteht aus zwei Strecken.

Weiter ergibt sich folgende

Stetigkeit: Es gilt für $t_1 < t_2 < t_2^*$ und $t_2 \to t_2^* - 0$

$$q(t_1, t_2) \nearrow q(t_1, t_2^*). \tag{6}$$

Denn aus

$$\lim_{t_2 \to t_2^* - 0} q(t_1, t_2) < q(t_1, t_2^*)$$

ergäbe sich ein Widerspruch mit Konvergenzsätzen [11] bei quasikonformen Abbildungen: Aus den Spiegelungen an den Bögen $t_1 \ldots t_2$ folgte in der Grenze eine Spiegelung mit einer Dilatationsschranke $< q(t_1, t_2^*)$ am Bogen $t_1 \ldots t_2^*$.

Folgerung: Ist $\mathfrak{r}$ eine geschlossene Jordankurve, dann gilt

$$q(t_1, t_2) \nearrow q_{\mathfrak{r}} \tag{7}$$

bei $t_1 = a$ für $t_2 \to b - 0$.

Wenn also aus einer geschlossenen Jordankurve ein im Durchmesser nach 0 strebendes Stückchen entfernt wird, strebt $q_{\mathfrak{r}}$ für den verbleibenden Jordanbogen gegen $q_{\mathfrak{r}}$ der vollen geschlossenen Jordankurve.

Natürlich gilt nicht immer das Gegenstück zu (6), $q(t_1, t_2) \searrow q(t_1, t_2^*)$ für $t_2 \to t_2^* + 0$; man vergleiche den Fall, $\mathfrak{r}$ besteht aus zwei Strecken.

Im Anschluß an diese einfachen Betrachtungen stellt sich die Frage nach dem Verhalten von $q(t_1, t_2)$ für $t_1 \to t_0 - 0$, $t_2 \to t_0 + 0$ bei fixiertem t_0 mit $t_1 < t_0 < t_2$, d.h. nach dem Verhalten von $q_{\mathfrak{r}}$ für auf einen Punkt schrumpfende Jordanbögen $\mathfrak{r}$. Daß nicht immer $q(t_1, t_2) \to 0$ gilt, zeigt wieder der Fall, $\mathfrak{r}$ besteht aus zwei Strecken (wobei der Parameterwert t_0 dem Knickpunkt entspricht). Jedoch gilt tatsächlich $q(t_1, t_2) \to 0$ im Falle eines analytischen Jordanbogens. (Es ist eine interessante offene Frage, unter welchen schwächeren Voraussetzungen dies ebenfalls gilt.) Dann kann man diese Aussage sogar noch präzisieren. Das ist der Inhalt von Satz 1, dessen Beweis wir uns nun zuwenden.

§3   Beweis von Satz 1

Dieser besteht im ersten Teile im Nachweis der Ungleichung

$$q_{\mathfrak{r}} \geq \frac{1}{12} |dk/ds| \cdot \varepsilon^2 + 0(\varepsilon^3). \tag{8}$$

Dabei spielt eine zentrale Rolle eine Koeffizientenabschätzung für quasikonform fortsetzbare schlichte konforme Abbildungen. Im zweiten Teile des Beweises wird noch explizit eine Spiegelung konstruiert, für die der Betrag der komplexen Dilatation $\leq$ der rechten Seite von (8) ist.

Zum *1. Beweisteil* vergleiche man Figur 1. Es wird eine Kette von schlichten konformen Abbildungen durchgeführt.

a.) *Übergang $z \to \mathfrak{z}$*: Erfolgt als eine lineare Transformation mit $z_0 \to \mathfrak{z}_0 = 0$, wobei $|d\mathfrak{z}/dz| = 1$ für $z = z_0$ und der Krümmungskreis an $\mathfrak{r}$ in z_0 in die reelle Achse übergeht. Das Bild $\mathfrak{r}_{\mathfrak{z}}$ von $\mathfrak{r}$ in der $\mathfrak{z} = \mathfrak{x} + i\mathfrak{y}$-Ebene hat dann in Umgebung von $\mathfrak{x} = 0$ die Darstellung

$$\mathfrak{z} = \mathfrak{x} + i\left(A\mathfrak{x}^3 + O(\mathfrak{x}^4)\right) \tag{9}$$

mit einer Konstanten A. Dabei können wir hinfort $A \geq 0$ annehmen (sonst nehmen wir zuvor eine Spiegelung vor). Dabei bezeichnet O jeweils hier und im folgenden das Landau-Symbol und zwar stets eine reelle Funktion. Falls wir dabei eine ganz bestimmte Funktion meinen, hängen wir an O noch einen Index an.

Zur Bestimmung von A benutzen wir die bekannte Darstellung

$$dk/ds = |dz/dt|^{-2} \cdot \Im\{z(t), t\}$$

(vgl. z.B. [12], [4], S.97ff.) für die Krümmung k der Kurve $\mathfrak{r}$ in Parameterdarstellung $z = z(t)$, wobei s die Bogenlänge ist. Wählen wir speziell als Kurvenparameter $t = \mathfrak{x}$, ergint sich für $z = z_0$

$$dk/ds = |dz/d\mathfrak{z}|^{-2} \cdot \Im\{z, \mathfrak{x}\} = \Im\{\mathfrak{z}, \mathfrak{x}\} = 6A \qquad (10)$$

wegen $|dz/d\mathfrak{z}| = 1$ für $\mathfrak{z} = 0$ und wegen der Invarianz der Schwarzschen Ableitung bei der linearen Transformation $\mathfrak{z} = \mathfrak{z}(z)$.

Für die Endpunkte $\mathfrak{z}_1$ and $\mathfrak{z}_2$ des Bildbogens von $\mathfrak{r}_\mathfrak{z}$ in der $\mathfrak{z}$-Ebene haben wir nach (9)

$$\mathfrak{z}_1 - \mathfrak{z}_2 = 2\varepsilon + 2\varepsilon O_1(\varepsilon) + 2i \cdot \left(A\varepsilon^3 + O(\epsilon^4)\right),$$

$$\mathfrak{z}_1 + \mathfrak{z}_2 = 2\varepsilon O_2(\varepsilon) + i \cdot O(\varepsilon^4).$$

b.) *Übergang* $\mathfrak{z} \to Z$: Erfolgt durch die ganz-lineare Transformation

$$Z = \frac{4}{\mathfrak{z}_1 - \mathfrak{z}_2}\mathfrak{z} - 2\frac{\mathfrak{z}_1 + \mathfrak{z}_2}{\mathfrak{z}_1 - \mathfrak{z}_2}, \qquad (11)$$

so daß der Bildbogen $\mathfrak{r}_Z$ in der Z-Ebene die Endpunkte ± 2 erhält. Dieser Bildbogen selbst hat die Gestalt

$$Z = 2\frac{\mathfrak{x} + i \cdot \left(A\mathfrak{x}^3 + O(\mathfrak{x}^4)\right) - \varepsilon \cdot O_2(\varepsilon) + i \cdot O(\varepsilon^4)}{\varepsilon + i \cdot A\varepsilon^3 + \varepsilon \cdot O_1(\varepsilon) + i \cdot O(\varepsilon^4)}, \qquad (12)$$

wieder mit $\mathfrak{x}$ als Kurvenparameter. Wir steigen nun vom Parameter $\mathfrak{x}$ um zu φ durch

$$\mathfrak{x} = (\varepsilon + \varepsilon \cdot O_1(\varepsilon)) \cdot \cos\varphi + \varepsilon \cdot O_2(\varepsilon), \quad 0 \le \varphi \le \pi. \qquad (13)$$

Die Endpunkte des Bildbogens erwischen wir jetzt durch $\varphi = 0$ und $\varphi = \pi$, wobei dort von $\mathfrak{x}$ der Wert $\Re\mathfrak{z}_1$ bzw. $\Re\mathfrak{z}_2$ von 2. Ordnung angenommen wird. Wir stellen uns noch zweckmäßig vor, für $\pi \le \varphi \le 2\pi$ entsteht das andere Ufer des Schlitzes bei $\mathfrak{r}$.

Nun wird in diesem neuen Parameter φ nach (12) der Bildbogen in der Z-Ebene gegeben durch

$$Z/2 = \frac{(1 + i \cdot A\varepsilon^2 \cos^2\varphi) \cdot \cos\varphi + O_1(\varepsilon) \cdot \cos\varphi + i \cdot O(\varepsilon^3)}{1 + i \cdot A\varepsilon^2 + O_1(\varepsilon) + i \cdot O(\varepsilon^3)},$$

wobei ab jetzt das Landau-Symbol O (ohne Index) auch zusätzlich von φ abhängen kann, dabei immer als analytisch in ε und φ vorzustellen ist. Es ist $O(\varepsilon^3)/\varepsilon^3$ für $\epsilon \to 0$ beschränkt und zwar gleichmäßig in φ. Wir rechnen noch weiter um

$$Z/2 = \left[(1 + i \cdot A\varepsilon^2 \cos^2 \varphi) \cdot \cos \varphi + O_1(\varepsilon) \cdot \cos \varphi + i \cdot O(\varepsilon^3)\right] \cdot$$
$$\cdot \left[1 - i \cdot A\varepsilon^2 + O_1(\varepsilon) + i \cdot O(\varepsilon^3) + O_1(\varepsilon) \cdot O_1(\varepsilon) + O(\varepsilon^3)\right], \qquad (14)$$
$$Z/2 = (1 - i \cdot A\varepsilon^2 \sin^2 \varphi) \cdot \cos \varphi + O(\varepsilon^3) + i \cdot O(\varepsilon^3).$$

Hier muß $O(\varepsilon^3) + i \cdot O(\varepsilon^3)$ für $\varphi = 0$ und $\varphi = \pi$ von 2. Ordnung verschwinden, so daß also diese Größe nach Division durch $\varepsilon^3 \cdot \sin^2 \varphi$ beschränkt sein muß. Dies bringt

$$(Z/2)^2 - 1 = -\sin^2 \varphi \cdot \left[1 + 2i \cdot A\varepsilon^2 \cos^2 \varphi + O(\varepsilon^3) + i \cdot O(\varepsilon^3)\right],$$
$$\left[(Z/2)^2 - 1\right]^{1/2} = i \cdot \sin \varphi \cdot \left[1 + i \cdot A\varepsilon^2 \cos^2 \varphi + O(\varepsilon^3) + i \cdot O(\varepsilon^3)\right]. \qquad (15)$$

c.) *Übergang $Z \to w$*: Erfolgt durch

$$Z = w + 1/w, \quad w = \frac{Z}{2} + \sqrt{\frac{Z^2}{4} - 1}. \qquad (16)$$

Mit (14) und (15) wird die Bildkurve $\mathfrak{C}_w$ (eine geschlossene analytische Jordankurve) in der w-Ebene dargestellt durch

$$w = e^{i\varphi} \cdot \left(1 - \frac{1}{2} A\varepsilon^2 \sin 2\varphi\right) + O(\varepsilon^3) + i \cdot O(\varepsilon^3).$$

d.) *Übergang $w \to \mathfrak{w}$*: Erfolgt durch Drehung $\mathfrak{w} = e^{i\pi/4} \cdot w$. Wir führen gleich noch eine harmlose Parametersubstitution durch: $\Psi = \varphi + \pi/4$. Die Bildkurve $\mathfrak{C}_{\mathfrak{w}}$ der $\mathfrak{w}$-Ebene ist dann gegeben durch

$$\mathfrak{w} = e^{i\Psi} \cdot \left(1 + \frac{1}{2} A\varepsilon^2 \cos 2\Psi\right) + O(\varepsilon^3) + i \cdot O(\varepsilon^3).$$

e.) *Übergang $\mathfrak{w} \to W$*: Erfolgt durch

$$\mathfrak{w} = W + \frac{1}{2} A\varepsilon^2 \frac{1}{W}, \quad W = \mathfrak{w} - \frac{1}{2} A\varepsilon^2 \frac{1}{\mathfrak{w}} + \frac{\cdots}{\mathfrak{w}^3} + \dots. \qquad (17)$$

Die Bildkurve $\mathfrak{C}_W$ in der W-Ebene hat dann die Darstellung

$$W = e^{i\Psi} \left(1 + \frac{1}{2} i A\varepsilon^2 \sin 2\Psi\right) + O(\varepsilon^3) + i \cdot (\varepsilon^3).$$

Durch die weitere Substitution $\chi = \Psi + \frac{1}{2}A\varepsilon^2 \sin 2\Psi$ erhalten wir

$$W = e^{i\chi} + O(\varepsilon^3) + i \cdot O(\varepsilon^3), \tag{18}$$

mit wieder in ε und χ analytischem $O(\varepsilon^3)$, wobei $O(\varepsilon^3)/\varepsilon^3$ für $\varepsilon \to 0$ in χ gleichmäßig beschränkt ist. Wegen

$$|W| = 1 + O(\varepsilon^3) \tag{19}$$

haben wir in der W-Ebene eine Annäherung an den Einheitskreis von 3. Ordnung, d.h. in der w- und $\mathfrak{w}$-Ebene Annäherung an gewisse Ellipsen von 3. Ordnung. Das Äußere dieser Kurven hat einen konformen Radius R der Form

$$R = 1 + O(\varepsilon^3). \tag{20}$$

f.) *Übergang* $W \to \mathfrak{W}$: Erfolgt durch schlichte konforme Abbildung des Äußeren von $\mathfrak{C}_W$ auf das Äußere des Einheitskreises, wobei in $\mathfrak{W} = \infty$ der Entwicklungstypus

$$\frac{W}{R} = \mathfrak{W} + \mathfrak{A}_0 + \frac{\mathfrak{A}_1}{\mathfrak{W}} + \frac{\mathfrak{A}_2}{\mathfrak{W}^2} + \ldots = G(\mathfrak{W}) \tag{21}$$

vorliegt. Es gilt

$$\log[G(\mathfrak{W})/\mathfrak{W}] = \mathfrak{A}_0 \cdot \mathfrak{W}^{-1} + \left(\mathfrak{A}_1 - \frac{1}{2}\mathfrak{A}_0^2\right) \cdot \mathfrak{W}^{-2} + \cdots,$$

$$\mathfrak{A}_0 = \frac{1}{2\pi i} \int\limits_{|\mathfrak{W}|=1} \log[G(\mathfrak{W})/\mathfrak{W}]d\mathfrak{W}. \tag{22}$$

Daneben gilt

$$\int\limits_{|\mathfrak{W}|=1} \log[G(\mathfrak{W})/\mathfrak{W}]\overline{d\mathfrak{W}} = - \int\limits_{|\mathfrak{W}|=1} \log[G(\mathfrak{W})/\mathfrak{W}]\frac{d\mathfrak{W}}{\mathfrak{W}^2} = 0.$$

Addieren wir das Konjugierte und durch $2\pi i$ Dividierte hiervon zu (22), entsteht

$$\mathfrak{A}_0 = \frac{1}{\pi i} \int\limits_{|\mathfrak{W}|=1} \log|G(\mathfrak{W})/\mathfrak{W}|d\mathfrak{W},$$

analog

$$\mathfrak{A}_1 - \frac{1}{2}\mathfrak{A}_0^2 = \frac{1}{\pi i} \int\limits_{|\mathfrak{W}|=1} \mathfrak{W} \cdot \log|G(\mathfrak{W})/\mathfrak{W}|d\mathfrak{W},$$

$$|\mathfrak{A}_0| \le O(\varepsilon^3), \quad |\mathfrak{A}_1| \le O(\varepsilon^3). \tag{23}$$

Nun haben wir in $\mathfrak{W} = \infty$ die Entwicklung

$$\frac{\mathfrak{w}}{R} = \frac{W}{R} + \frac{A\varepsilon^2}{2R^2} \cdot \frac{R}{W} = \mathfrak{W} + \frac{\mathfrak{A}_1 + A\varepsilon^2 R^{-2}/2}{\mathfrak{W}} + \frac{\cdots}{\mathfrak{W}^2} + \cdots + \mathrm{const.} \quad (24)$$

Aus der $Q_{\mathfrak{r}}$-quasikonformen Spiegelung an $\mathfrak{r}$ entsteht nun — wie in der Einleitung gesagt — eine $Q_{\mathfrak{r}}$-quasikonforme Spiegelung an der "ellipsennahen" Kurve der $\mathfrak{w}$-Ebene, und diese induziert in bekannter Weise (vgl. z.B. [9], §3) eine $Q_{\mathfrak{r}}$-quasikonforme Fortsetzung der schlichten konformen und hydrodynamisch normierten Abbildung des Äußeren des Einheitskreises der $\mathfrak{W}$-Ebene auf das Äußere der Kurve der Ebene von $\mathfrak{w}/R$. Dann können wir — das ist der entscheidende letzte Schritt dieser Schlußkette — in (24) den Koeffizienten von $1/\mathfrak{W}$ nach [5] abschätzen:

$$\left| \mathfrak{A}_1 + \frac{1}{2}A\varepsilon^2 R^{-2} \right| \leq q_{\mathfrak{r}},$$

oder mit (20) und (23)

$$q_{\mathfrak{r}} \geq \frac{1}{2}A\varepsilon^2 + O(\varepsilon^3).$$

Das ist im Vergleich mit (10) die Abschätzung (8), d.h. das Ziel des ersten Beweisteiles.

Im *2. Beweisteil* zu Satz 1 konstruieren wir effektiv an der geschlossenen Kurve $\mathfrak{C}_{\mathfrak{w}}$ der $\mathfrak{w}$-Ebene (vgl. wieder Figur 1) eine quasikonforme Spiegelung mit einem Betrag der komplexen Dilatation, der $\leq \frac{1}{2}A\varepsilon^2 + O(\varepsilon^3)$ ist.

Dazu zeichnen wir innerhalb des Bildes $\mathfrak{C}_W$ in der W-Ebene eine zu $W = 0$ konzentrische Kreislinie eines Radius' $R^* = 1 + O(\varepsilon^3)$, was nach (19) möglich ist. Nun nehmen wir die folgende Fortsetzung der Abbildung (17) zu Hilfe:

$$\mathfrak{w} = \begin{cases} W + \frac{1}{2}A\varepsilon^2/W & \text{für } |W| \geq R^*, \\ W + \frac{1}{2}A\varepsilon^2 R^{*-2}\bar{W} & \text{für } |W| \leq R^*, \end{cases} \quad (25)$$

deren komplexe Dilatation $\frac{1}{2}A\varepsilon^2 R^{*-2} = \frac{1}{2}A\varepsilon^2 + O(\varepsilon^5)$ ist. Zur Herstellung der angekündigten Spiegelung an $\mathfrak{C}_{\mathfrak{w}}$ bilden wir zunächst das Äußere von $\mathfrak{C}_{\mathfrak{w}}$ gemäß (25) konform auf's Äußere von $\mathfrak{C}_W$ ab. Nun spiegeln wir an der Bildkurve $\mathfrak{C}_W$ durch

$$W^* = r^2(\arg W)/\bar{W}, \quad (26)$$

wenn $r = r(\vartheta)$ die Darstellung von $\mathfrak{C}_W$ in Polarkoordinaten r, ϑ ist. Bei dieser "Ahlforsschen Spiegelung" [1] (vgl. auch [2], S.36) an $\mathfrak{C}_W$ ist der Betrag der komplexen Dilatation $\leq \max |\cos\alpha|$, wenn α den Winkel zwischen den Strahlen $\arg W = \mathrm{const}$ und den Tangenten an $\mathfrak{C}_W$ bezeichnet. Wegen dem bei (18) Gesagten (zur Analytizität von $\mathfrak{C}_W$ in ε und χ) gilt $\max |\cos\alpha| \leq O(\varepsilon^3)$.

Wenn wir daher nun das Innere von $\mathfrak{C}_W$ durch die Umkehrung von (25) auf's Innere von $\mathfrak{C}_\mathfrak{W}$ abbilden, erhalten wir insgesamt durch Zusammensetzung die gewünschte Spiegelung an $\mathfrak{C}_\mathfrak{W}$.

Der Beweis von Satz 1 ist damit zu Ende.

§4   Asymptotisches Verhalten der möglichst konformen Spiegelung an $\mathfrak{r}$

a.) Vorbereitend stellen wir einige Ergebnisse aus [8], [9] (in verschärfter Form) zusammen.

Es sei $f(\mathfrak{W}) = \mathfrak{W} + a_1/\mathfrak{W} + a_2/\mathfrak{W}^2 + \cdots \not\equiv \mathfrak{W}$ eine schlichte konforme Abbildung von $|\mathfrak{W}| > \rho$ mit $0 < \rho < 1$. Als Bild von $|\mathfrak{W}| = 1$ entsteht eine geschlossene analytische Jordankurve $\mathfrak{C}$. Die Grunskyschen Koeffizienten $C_{kl} = \sqrt{kl} \cdot a_{kl}$ werden wie üblich der Entwicklung

$$-\log \frac{f(\mathfrak{W}) - f(\mathfrak{W}')}{\mathfrak{W} - \mathfrak{W}'} = \sum_{k,l=1}^{\infty} a_{kl} \cdot \mathfrak{W}^{-k}\mathfrak{W}'^{-l}$$

entnommen. Dann gilt mit dem reziproken Fredholmschen Eigenwert κ ($0 < \kappa < 1$) von $\mathfrak{C}$ nach [6], [14] für alle Systeme komplexer Zahlen x_k stets

$$\left| \sum_{k,l=1}^{\infty} C_{kl}x_k x_l \right| \le \kappa \cdot \sum_{k=1}^{\infty} |x_k|^2,$$

falls letztere Reihe konvergiert. Für ein (existierendes) System $\{x_k\}$ mit

$$\sum_{k=1}^{\infty} |x_k|^2 = 1 \quad \text{und} \quad \sum_{k,l=1}^{\infty} C_{kl}x_k x_l = \kappa \tag{27}$$

(letztere Reihe konvergiert absolut) gilt [8]

$$\kappa x_l = \sum_{k=1}^{\infty} \overline{C_{kl}x_k}, \qquad l = 1, 2, \ldots . \tag{28}$$

Wir setzen noch $C_{kl} = C_{kl}^* \cdot \rho^{k+l}$, wobei C_{kl}^* die Grunskyschen Koeffizienten für die Abbildung $f(\rho\mathfrak{W})/\rho$ sind, für die nach [13] (S.59) gilt

$$\sum_{k=1}^{\infty} |C_{kl}^*|^2 \le 1. \tag{29}$$

Aus (27), (28) folgt nun mit (29)

$$\kappa|x_l| \le \sum_{k=1}^{\infty} |C_{kl}x_k| = \sum_{k=1}^{\infty} |C_{kl}^*|\cdot|x_k|\cdot\rho^{k+l} \le \rho^l\cdot\left[\sum_{k=1}^{\infty} |x_k|^2\rho^{2k}\right]^{1/2} \le \rho^{l+1}. \quad (30)$$

Nach diesen Vorbereitungen können wir folgende Verschärfung von Satz 4 in [9] beweisen.

Satz 5. *Die möglichst konforme Fortsetzung der nur für $|\mathfrak{w}| > 1$ betrachteten schlichten konformen Abbildung $f(\mathfrak{w})$ nach $|\mathfrak{w}| < 1$ und damit die möglichst konforme Spiegelung an $\mathfrak{C}$ ist durch ein quadratisches Differential ohne jegliche Nullstellen innerhalb und auf $\mathfrak{C}$ zu beschreiben, falls mit*

$$\Phi(\rho) = \left[\left(\sum_{n=0}^{\infty} \sqrt{n+2}\,\rho^n\right)^2 + (1-\rho^2)^{-1}\right]\rho^6 \quad (31)$$

gilt

$$\Phi(\rho) < \kappa^2. \quad (32)$$

Insbesondere gilt unter der Voraussetzung (32) auch $q_{\mathfrak{C}} = \kappa$.

Beweis. Für die wie in Satz 7 in [7] oder (13) in [9] gebildete komplexe Eigenfunktion

$$F(\mathfrak{w}) = i\sum_{n=1}^{\infty} \overline{x_n}\,\mathfrak{w}^{-n}/\sqrt{n} \quad (|\mathfrak{w}| \ge 1) \quad (33)$$

liegt Schlichtheit für $|\mathfrak{w}| \ge 1$ (vgl. [13], S.44) sicher dann vor, was nach Satz 7 in [7] unseren Satz 5 nach sich zieht, wenn

$$\sum_{n=2}^{\infty} \sqrt{n}\,|x_n| < |x_1|. \quad (34)$$

Aus (30) bekommen wir

$$\sum_{n=2}^{\infty} \sqrt{n}\,|x_n| \le \kappa^{-1}\cdot\sum_{n=2}^{\infty} \sqrt{n}\,\rho^{n+1}, \quad \sum_{n=2}^{\infty} |x_n|^2 \le \kappa^{-2}\sum_{n=2}^{\infty} \rho^{2n+2}. \quad (35)$$

Damit ist (34) sicher erfüllt, wenn

$$\kappa^{-1}\sum_{n=2}^{\infty} \sqrt{n}\,\rho^{n+1} < \left[1 - \kappa^{-2}\sum_{n=2}^{\infty} \rho^{2n+2}\right]^{1/2} \le \left[1 - \sum_{n=2}^{\infty} |x_n|^2\right]^{1/2} = |x_1|.$$

Dies liefert (32). Die letzte Zeile der Behauptung von Satz 5 folgt noch aus Satz 2 in [9]. $\square$

Bemerkungen. 1.) Wegen der auf M. Schiffer zurückgehenden Ungleichung $\kappa \le \rho^2$ (vgl. [7]) kann (32) höchstens erfüllbar sein, wenn für die (monoton steigende) Funktion $\Phi(\rho)/\rho^4$ gilt $\Phi(\rho)/\rho^4 < 1$, was zu $\rho < 0,367\ldots$ führt. Für größere ρ ist Satz 5 also wertlos.

2.) Es gilt $\Phi(\rho)/\rho^6 \ge 3$, weil $\Phi(\rho)/\rho^6$ auch monoton steigend ist und den Grenzwert 3 für $\rho \to 0$ besitzt.

3.) Man kann durch weitere geringe Vergröberung die Bedingung (32) noch in eine leichter nachprüfbare Form bringen, die nicht mehr die (im allgemeinen schwer bestimmbare) Größe κ enthält. Wegen $\sum n|a_n|^2 \le \kappa^2$ [6] ist nämlich statt (32) auch

$$\Phi(\rho) < \sum_{n=1}^{\infty} n|a_n|^2 \tag{36}$$

hinreichend für Gleichheit in der Ungleichung $\kappa \le q_{\mathfrak{C}}$, oder auch einfach $\Phi(\rho) < |a_1|^2$.

4.) Beiläufig bemerkt, läßt sich (14) in [8] zu

$$0 \le \kappa - \kappa_n \le \frac{2\rho^{2n+4}}{\kappa\sqrt{1-\rho^4}} \tag{37}$$

verschärfen. Dazu schätzen wir nach (15) in [8] so weiter ab mit (30)

$$0 \le \kappa - \kappa_n \le 2\left[\rho^2 \sum_{k=1}^{\infty} |x_k|^2 \cdot \sum_{k=n+1}^{\infty} \left|\rho^k \frac{1}{\kappa}\rho^{k+1}\right|^2\right]^{1/2} .$$

Im folgenden entspricht $f(\mathfrak{W})$ bis auf einen harmlosen Normierungsfaktor der Form $1 + O(\varepsilon^3)$ (vgl. (20)) der Abbildung $\mathfrak{W} \to \mathfrak{w}$ (vgl. Figur 1).

b.) Der *Beweis von Satz 2* ergibt sich jetzt so. Für hinreichend kleine ε gilt nach (24) für den Koeffizienten von $1/\mathfrak{W}$

$$a_1 = \frac{1}{2}A\varepsilon^2 + O(\varepsilon^3) + iO(\varepsilon^3), \quad |a_1|^2 = \frac{1}{4}A^2\varepsilon^4 + O(\varepsilon^5). \tag{38}$$

Andererseits ist $\mathfrak{w}(\mathfrak{W})$ nach $|\mathfrak{W}| > \rho$ schlicht und analytisch fortsetzbar, wobei wählbar ist

$$\rho = \text{const} \cdot \varepsilon \tag{39}$$

mit einer von ε unabhängigen Konstanten. Denn an $\mathbf{r}_Z$ (vgl. Figur 1) läßt sich in einem Umgebungsstreifen einer Breite der Form $\mathrm{const}/\varepsilon$ spiegeln, und diese Spiegelung gibt in bekannter Weise Anlaß zu einer schlichten und analytischen Fortsetzung von $\mathbf{u}(\mathfrak{W})$ nach $|\mathfrak{W}| > \mathrm{const} \cdot \varepsilon$.

Aus (38) und (39) ergeben sich Satz 2 und 3 nach Satz 5 (dort insbesondere Bemerkung 2. und 3.) wegen $\Phi(\rho) = \mathrm{const} \cdot \varepsilon^6 + O(\varepsilon^7)$.

c.) *Beweis von Satz 4.* Im Anschluß an die vorigen Überlegungen ergibt sich sogar

$$\sum_{n=2}^{\infty} \sqrt{n} \cdot |x_n| \cdot |\mathfrak{W}|^{-n-1} \leq c \cdot \varepsilon \cdot |x_1| \cdot |\mathfrak{W}|^{-2} \quad \text{für} \quad |\mathfrak{W}| \geq 1 \qquad (40)$$

mit einer Konstanten c und für hinreichend kleines ε (bei fester Trägerkurve für $\mathbf{r}$). Denn (40) folgt aus

$$\sum_{n=2}^{\infty} \sqrt{n}|x_n| \leq c \cdot \varepsilon \cdot |x_1|$$

und dies wegen (35) aus

$$\left(\sum_{n=2}^{\infty} \sqrt{n}\rho^{n+1} \right)^2 + c^2\varepsilon^2 \frac{\rho^6}{1-\rho^2} \leq c^2\varepsilon^2\kappa^2,$$

was für kleine ε sicherlich mit einer geeigneten Konstanten c richtig ist wegen $\kappa \geq |a_1|$ und (38), (39).

Nun gilt weiter nach (28)

$$\kappa x_1 - \bar{a}_1 \bar{x}_1 = \sum_{k=2}^{\infty} \overline{C_{k1} x_k}.$$

Hier wird wegen $|C_{k1}| \leq \rho^{k+1}$ (vgl. (29)) und (30) die Ungleichung

$$\left| \sum_{k=2}^{\infty} C_{k1} x_k \right| \leq \frac{1}{\kappa} \sum_{k=2}^{\infty} \rho^{2k+2} = \frac{1}{\kappa} \frac{\rho^6}{1-\rho^2}$$

befriedigt, so daß sich wegen $\kappa = q_{\mathbf{r}} = \frac{1}{2} A \varepsilon^2 + O(\varepsilon^3)$ und (38), (39) ergibt

$$\mathfrak{Im} x_1 = O(\varepsilon). \qquad (41)$$

Weiter gilt nach (35), (39)

$$\sum_{k=2}^{\infty} |x_k|^2 \leq \frac{1}{\kappa^2} \frac{\rho^6}{1-\rho^2} = O(\varepsilon^2), \quad |x_1|^2 = 1 - \sum_{k=2}^{\infty} |x_k|^2 = 1 - O(\varepsilon^2),$$

so daß folgt

$$x_1 = 1 - O(\varepsilon^2) + i \cdot O(\varepsilon). \tag{42}$$

Nach (40) gilt für die Ableitung der komplexen Eigenfunktion (33)

$$\frac{d}{d\mathbf{w}} F(\mathbf{w}) = -i \cdot \mathbf{w}^{-2} \bar{x}_1 \left(1 + c\varepsilon F^*(\mathbf{w}, \varepsilon)\right) \tag{43}$$

mit $|F^*(\mathbf{w}, \varepsilon)| \leq 1$ für $|\mathbf{w}| \geq 1$ und alle hinreichend kleinen ε.

Das beweist im Verein mit (42) Satz 4, da die Schar S — in die $\mathbf{w}$-Ebene überpflanzt — durch $\mathfrak{Im}(i\mathbf{w}^{-1}) = $ const charakterisiert wird, die eigentliche Schar der Hauptverzerrungsrichtungen (Richtungen der großen Achsen der infinitesimalen Ellipsen, die Bilder infinitesimaler Kreise) durch $\mathfrak{Im}\,dF = 0$.

Schrifttum

[1] Ahlfors, L.V., Remarks on the Neumann-Poincaré integral equation. Pacific J. Math. *2*, 271–280 (1952).

[2] Gaier, D., Konstruktive Methoden der konformen Abbildung. Springer-Verlag, Berlin-Göttingen-Heidelberg 1964.

[3] Gehring, F.W., Characteristic properties of quasidisks. Les Presses de l'Université de Montréal, Montréal 1982.

[4] König, R. und K.H. Weise, Mathematische Grundlagen der höheren Geodäsie und Kartographie, I. Band: Das Erdsphäroid und seine konformen Abbildungen. Springer-Verlag, Berlin-Göttingen-Heidelberg 1951.

[5] Kühnau, R., Wertannahmeprobleme bei quasikonformen Abbildungen mit ortsabhängiger Dilatationsbeschränkung. Math. Nachr. *40*, 1–11 (1969).

[6] —, Zu den Grunskyschen Coeffizientenbedingungen. Ann. Acad. Sci. Fenn. Ser. A. I. Math. *6*, 125–130 (1981).

[7] —, Quasikonforme Fortsetzbarkeit, Fredholmsche Eigenwerte und Grunskysche Koeffizientenbedingungen. Ann. Acad. Sci. Fenn. Ser. A. I. Math. *7*, 383–391 (1982).

[8] —, Zur Berechnung der Fredholmschen Eigenwerte ebener Kurven. ZAMM *66*, 193–200 (1986).

[9] —, Wann sind die Grunskyschen Koeffizientenbedingungen hinreichend für Q-quasikonforme Fortsetzbarkeit? Comment. Math. Helv. *61*, 290–307 (1986).

[10] —, Möglichst konforme Spiegelung an einer Jordankurve. Jahresber. DMV *90*, 90–109 (1988).

[11] Lehto, O. und K.I. Virtanen, Quasikonforme Abbildungen. Springer-Verlag, Berlin-Heidelberg-New York 1965.

[12] Pick, G., Zur Theorie der konformen Abbildung kreisförmiger Bereiche. Rend. Circ. Mat. Palermo *37*, 341–344 (1914).

[13] Pommerenke, Chr., Univalent functions. Vandenhoeck & Ruprecht, Göttingen 1975.

[14] Schiffer, M., Fredholm eigenvalues and Grunsky matrices. Ann. Polon. Math. *39*, 149–164 (1981).

[15] Schober, G., Estimates for Fredholm eigenvalues based on quasiconformal mapping. Lect. Notes Math. *333*, 211–217, Springer-Verlag, 1973.

[16] Strebel, K., On the existence of extremal Teichmueller mappings. J. d'Analyse Math. *30*, 464–480 (1976).

Reiner Kühnau
Sektion Mathematik der
Martin-Luther-Univ. Halle-Wittenberg
DDR-402 Halle an der Saale

Heinz Leutwiler

On BMO and the Torsion Function

Introduction

The space BMO has been extensively studied by many authors (see [6] for a good exposition of this topic). However, whereas the real-variable theory is highly developed in any dimension, its counterpart, the space BMOH of *harmonic* functions of *bounded mean oscillation* seems not to be well understood in case the dimension is greater than two. In $I\!\!R^2_+$, BMOH agrees with the real part of BMOA, the space of analytic functions of bounded mean oscillation, and hence a lot is knwon (see [1], for an expository article). It is the purpose of this paper to extend some of these results to higher dimensions.

In §1 we recall some of the known facts concerning BMOH, needed in the sequel. Then, in §2 we prove an invariant version of the John-Nirenberg Theorem for the upper half space $I\!\!R^n_+$, using harmonic measure. As an application of this result, in §3, we shall give several characterizations of BMOH $(I\!\!R^n_+)$. In §4 we show that every harmonic function of class BMOH $(I\!\!R^n_+)$ is Lipschitz-continuous with respect to the hyperbolic distance, and hence BMOH $(I\!\!R^n_+) \subset$ BLOCH $(I\!\!R^n_+)$. Finally, in §5, we extend the concept of a BMO-domain from $\mathbb{C}$ to $I\!\!R^n (n = 2)$ and list several characterizations of such domains. In particular we shall stress the close connection of a BMO-domain with the *torsion function*, a concept studied by J. Hersch [8].

§1   Harmonic functions of bounded mean oscillation

Recall the classical definition: A function $f \in L^1_{loc}(\mathbb{R}^n)$ $(n = 1)$ belongs to the space BMO $(\mathbb{R}^n)$, provided

$$\| f \|_* = \sup_Q \frac{1}{|Q|} \int_Q |f(x) - f_Q| d^n x < \infty, \tag{1.1}$$

where $|Q|$ denotes the volume of the cube $Q \subset \mathbb{R}^n$ and

$$f_Q = \frac{1}{|Q|} \int_Q f \, d^n x \tag{1.2}$$

its average over Q. The supremum ranges, by definition, over all cubes Q in $\mathbb{R}^n$, parallel to the coordinate axes.

If $f \in$ BMO $(\mathbb{R}^{n-1})$ $(n = 2)$ then f is integrable on $\mathbb{R}^{n-1}$ with respect to the measure $d^{n-1}y/(1 + \| y \|^n)$, where $y = (y_1, \ldots, y_{n-1})$ and $\| y \| = (\sum_1^{n-1} y_i^2)^{1/2}$. Consequently its Poisson integral

$$h(x) = \int_{\mathbb{R}^{n-1}} f(y) P_x(y) d^{n-1} y, \tag{1.3}$$

where

$$P_x(y) = \frac{\kappa x_n}{(\sum_1^{n-1} (x_i - y_i)^2 + x_n^2)^{n/2}} \tag{1.4}$$

and

$$\kappa = \kappa_n = \pi^{-n/2} \Gamma\left(\frac{n}{2}\right), \tag{1.5}$$

is well defined and represents a harmonic function on the upper half space $\mathbb{R}^n_+ = \{(x_1, \ldots, x_n) \in \mathbb{R}^n : x_n > o\}$.

The following result, due to Fefferman-Stein [5], is crucial:

Lemma 1.1. *Let $f \in$ BMO $(\mathbb{R}^{n-1})$. To each $x = (x', x_n) \in \mathbb{R}^n_+$ associate the cube Q_x, parallel to the coordinate axes, with midpoint $x' \in \mathbb{R}^{n-1}$ and sides of length $2x_n$. Then there is a constant $d > 0$, depending only on n, such that*

$$\int_{\mathbb{R}^{n-1}} |f(y) - f_{Q_x}| P_x(y) \, d^{n-1} y \leqq d \, \| f \|_*, \tag{1.6}$$

for all $x \in \mathbb{R}^n_+$.

Once this lemma has been established it is easy to verify

Theorem 1.2. *Let $f \in \mathrm{BMO}\,(I\!\!R^{n-1})$ and h denote its Poisson integral* (1.3). *Then*

$$\| h \|_* = \sup_{x \in I\!\!R^n_+} \int_{I\!\!R^{n-1}} |f(y) - h(x)| P_x(y)\, d^{n-1}y < \infty. \tag{1.7}$$

Conversely, if for some $f \in L^1(d^{n-1}y/(1+ \| y \|^n))$, the Poisson integral (1.3) *satisfies the condition* (1.7), *it is in class* $\mathrm{BMO}\,(I\!\!R^{n-1})$. *Furthermore, there is a constant $C > 0$, depending only on the dimension n, such that*

$$\frac{1}{C} \| f \|_* \leqq \| h \|_* \leqq C \| f \|_* . \tag{1.8}$$

For proofs of Lemma 1.1 and Theorem 1.2 we refer, e.g., to [6]. In order to avoid the boundary function f in (1.7) and extend at the same time the condition (1.7) to arbitrary domains in $I\!\!R^n$ we proceed as follows (see [10]). Let Ω denote a domain in $I\!\!R^n$ $(n = 2)$ and $H(\Omega)$ the set of harmonic functions on Ω. Assuming that the absolute value $|h|$ of $h \in H(\Omega)$ admits a harmonic majorant, it also admits a *least harmonic majorant*. In what follows we shall always denote this least harmonic majorant by $M|h|$, resp. $M_\Omega |h|$.

Note that if h admits a harmonic majorant, so does the function $|h - h(x)|$, where x denotes an arbitrary (fixed) point of Ω. Consequently $M|h - h(x)|$ is a well defined harmonic function on Ω. Evaluating it at the *same* point $x \in \Omega$ we obtain the function

$$x \to M|h - h(x)|(x), \tag{1.9}$$

defined on Ω.

Obviously, in the case of the upper half space $I\!\!R^n_+$, (1.9) is exactly the function occuring on the right hand side of (1.7). It therefore makes sense to introduce the following notion.

Definition. A harmonic function h on a domain $\Omega \subset I\!\!R^n$ $(n = 2)$ is called of bounded mean oscillation if $|h|$ admits a harmonic majorant, satisfying the condition

$$\| h \|_* = \sup_{x \in \Omega} M|h - h(x)|(x) < \infty. \tag{1.10}$$

Let us denote the class of harmonic functions of bounded mean oscillation on Ω by BMOH (Ω).

Then Theorem 1.1 says that BMOH $(I\!R^n_+)$ consists precisely of those harmonic functions which are representable as a Poisson integral of some boundary function of class BMO $(I\!R^{n-1})$.

Remark. In [10], where (X, H) had more generally been taken as an arbitrary Brelot space with $1 \in H(X)$, we simply wrote B (X) instead of BMOH (X). In addition $\| h \|_*$ as defined here differs from the same symbol defined there by a factor 2 (see (1.7) in [10]).

Identifying $h \in$ BMOH (Ω) with the functions $h+$ const., $\| h \|_*$ defines a norm on the space BMOH (Ω). In fact (BMOH $(\Omega)/I\!R$, $\| \, . \, \|_*$) is a Banach space ([10], Prop. 4.2). Note also that BMOH (Ω) contains all bounded harmonic functions on Ω; more precisely, we have (see also [10], Theorem 6.2, where equality has been characterized)

$$\| h \|_* \leqq \| h \|, \tag{1.11}$$

where $\| h \|$ denotes the sup-norm of h. On the other hand, every function $h \in$ BMOH (Ω) is quasi-bounded on Ω ([10], Prop. 2.1).

§2   Invariant form of the John-Nirenberg theorem

In [10] (Lemma 8.1) the following invariant form of the John-Nirenberg theorem had been considered.

Theorem 2.1. *Let* $f \in$ BMO$(I\!R^{n-1})$ *and* h *its Poisson integral on* $I\!R^n_+$ $(n = 2)$. *There are constants* $a, b > 0$, *depending only on* n, *such that for all* $\lambda > 0$ *and* $x \in I\!R^n_+$,

$$\mu_x\{y \in I\!R^{n-1} : |f(y) - h(x)| > \lambda\} \leqq ae^{-b\lambda/\|f\|_*}, \tag{2.1}$$

where μ_x *denotes the harmonic measure relative to the point* $x \in I\!R^n_+$.

The proof given in [10], however, is incorrect, as kindly pointed out by W. Hansen. We now take the opportunity to present a corrected version of this proof. A misprint in the citation of the John-Nirenberg theorem unfortunately caused the error. Correctly stated, it says the following (see [9]): There are positive constants b, c, depending only on n, such that for any function

$f \in \mathrm{BMO}\,(\mathbb{R}^{n-1})$ and any cube Q of $\mathbb{R}^{n-1}$, parallel to the coordinate axes, we have

$$|\{x \in Q : |f(x) - f_Q| > \lambda\}| \leqq |Q|\,ce^{-b\lambda/\|f\|_*} \qquad (2.2)$$

for all $\lambda > 0$, where $|\{\cdot\}|$ denotes Lebesgue measure of $\{\cdot\}$. An obvious adaptation to $\mathbb{R}^{n-1}$ of the proof given in [6] on page 232 (with $\alpha = 3/2$ replaced by $\alpha = 2$) shows that in fact we may always choose

$$c = \sqrt{2} \quad \text{and} \quad b = \log 2/2^{n+1}. \qquad (2.3)$$

On these grounds we now prove Theorem 2.1 in a way similar to the proof of Lemma 1.1. In particular we shall use the following elementary property: For any two cubes Q_1, $Q_2 \subset \mathbb{R}^{n-1}$ we have

$$Q_1 \subset Q_2 \Rightarrow |f_{Q_1} - f_{Q_2}| \leqq (|Q_2|/|Q_1|)\,\|\,f\,\|_* \,. \qquad (2.4)$$

Indeed,

$$|f_{Q_1} - f_{Q_2}| \leqq \frac{1}{|Q_1|} \int\limits_{Q_1} |f - f_{Q_2}| d^{n-1}x$$

$$\leqq \frac{|Q_2|}{|Q_1|} \cdot \frac{1}{|Q_2|} \int\limits_{Q_2} |f - f_{Q_2}| d^{n-1}x \leqq \frac{|Q_2|}{|Q_1|}\,\|\,f\,\|_* \,.$$

Proof of Theorem 2.1. Let $f \in \mathrm{BMO}\,(\mathbb{R}^{n-1})$ and h its Poisson integral. Fix a cube $Q = Q_x$, $x = (x', x_n) \in \mathbb{R}^n_+$, parallel to the coordinate axes, with midpoint $x' \in \mathbb{R}^{n-1}$ and sides of length $2x_n$. According to Lemma 1.1 we have

$$|h(x) - f_Q| \leqq M|h - f_Q|(x) \leqq d\,\|\,f\,\|_* \,.$$

Denoting by μ_x the harmonic measure with respect to $x \in \mathbb{R}^n_+$ we therefore conclude that

$$\left.\begin{aligned}
&\mu_x\{y \in \mathbb{R}^{n-1} : |f(y) - h(x)| > \lambda\} \\
\leqq\,&\mu_x\{y \in \mathbb{R}^{n-1} : |f(y) - f_Q| > \lambda - |f_Q - h(x)|\} \\
\leqq\,&\mu_x\{y \in \mathbb{R}^{n-1} : |f(y) - f_Q| > \lambda - d\,\|\,f\,\|_*\}.
\end{aligned}\right\} \qquad (2.5)$$

Thus the problem is reduced to estimate the distribution function

$$m(\lambda) = \mu_x\{y \in \mathbb{R}^{n-1} : |f(y) - f_{Q_x}| > \lambda\}.$$

In order to do so we introduce the cubes $Q_k = 2^k Q_x$ $(k = 0, 1, \ldots,)$ with sides of length $2^{k+1}x_n$.

Obviously, for $y \in Q_{k+1} \setminus Q_k$, the Poisson kernel (1.4) can be estimated from above by

$$P_x(y) \leqq \frac{\kappa}{2^{kn} x_n^{n-1}}.\tag{2.6}$$

Observing that for any Borel set $e \subset \mathbb{R}^{n-1}$ with characteristic function χ_e

$$\mu_x(e) = \int_{\mathbb{R}^{n-1}} \chi_e(y) P_x(y) d^{n-1}y$$

we conclude from (2.6) that

$$
\begin{aligned}
m(\lambda) = \ &\mu_x\{y \in Q_0 : |f(y) - f_{Q_0}| > \lambda\} \\
&+ \sum_0^\infty \mu_x\{y \in Q_{k+1} \setminus Q_k : |f(y) - f_{Q_0}| > \lambda\} \\
\leqq \ &\frac{\kappa}{x_n^{n-1}} |\{y \in Q_0 : |f(y) - f_{Q_0}| > \lambda\}| \\
&+ \kappa \sum_0^\infty 2^{-kn} x_n^{1-n} |\{y \in Q_{k+1} \setminus Q_k : |f(y) - f_{Q_0}| > \lambda\}| \\
\leqq \ &2^{n-1} \frac{\kappa}{|Q_0|} |\{y \in Q_0 : |f(y) - f_{Q_0}| > \lambda\}| \\
&+ \kappa \sum_0^\infty \frac{2^{2n-k-2}}{|Q_{k+1}|} |\{y \in Q_{k+1} \setminus Q_k : |f(y) - f_{Q_0}| > \lambda\}|.
\end{aligned}
$$

On account of (2.4)

$$|f_{Q_k} - f_{Q_0}| \leqq \sum_1^k |f_{Q_j} - f_{Q_{j-1}}| \leqq 2^{n-1} k \| f \|_* .$$

Consequently, by (2.2), we have

$$\frac{1}{|Q_k|} |\{y \in Q_k : |f(y) - f_{Q_0}| > \lambda\}|$$

$$\leqq \frac{1}{|Q_k|} |\{y \in Q_k : |f(y) - f_{Q_k}| > \lambda - |f_{Q_k} - f_{Q_0}|\}|$$

$$\leqq \frac{1}{|Q_k|} |\{y \in Q_k : |f(y) - f_{Q_k}| > \lambda - 2^{n-1} k \| f \|_*\}|$$

$$\leqq c e^{-b(\lambda - 2^{n-1} k \| f \|_*)/\|f\|_*} = c e^{2^{n-1} kb} e^{-b\lambda/\|f\|_*}.$$

Inserted in $m(\lambda)$ this yields

$$m(\lambda) \leqq 2^{n-1}\kappa c e^{-b\lambda/\|f\|_*} + 2^{2(n-1)}\kappa \sum_0^\infty 2^{-k} c e^{2^{n-1}(k+1)b} e^{-b\lambda/\|f\|_*}$$

$$= 2^{n-1} c\kappa \left(1 + 2^{n-1} e^{2^{n-1}b} \sum_0^\infty 2^{-k} e^{2^{n-1}kb} \right) e^{-b\lambda/\|f\|_*}.$$

By (2.3) we have $b = \frac{\log 2}{2 \cdot 2^n}$ and hence

$$\sum_0^\infty 2^{-k} e^{2^{n-1}kb} = \sum_0^\infty (2^{-3/4})^k = \frac{2^{3/4}}{2^{3/4} - 1}.$$

In view of (2.5) there results

$$\mu_x\{y \in I\!\!R^{n-1} : |f(y) - h(x)| > \lambda\}$$
$$\leqq \text{const } e^{-b(\lambda - d\|f\|_*)/\|f\|_*} = a e^{-b\lambda/\|f\|_*},$$

where a denotes a constant which only depends on n. $\Box$

§3 Characterizations of BMOH

As a consequence of Theorem 2.1 we get the following characterizations of BMOH $(I\!\!R_+^n)$ $(n = 2)$:

Theorem 3.1. *Let h be harmonic on $I\!\!R_+^n$. The following statements are equivalent:*

(i) $h \in \text{BOMH}(I\!\!R_+^n)$;

(ii) *for some real $p = 1$, $|h|^p$ admits a harmonic majorant and*

$$\| h \|_{*,p} = \left(\sup_{x \in I\!\!R_+^n} M|h - h(x)|^p(x) \right)^{1/p} < \infty; \qquad (3.1)$$

(iii) *for all real $p = 1$, $|h|^p$ admits a harmonic majorant and (3.1) holds;*

(iv) *h^2 admits a harmonic majorant and the Green potential*

$$\int_{I\!\!R_+^n} G(\cdot, z) \| \text{grad}\, h \|^2 (z) d^n z$$

is bounded on $\mathbb{R}^n_+$ (G = Green's function for $\mathbb{R}^n_+$);

(v) *h is the Poisson integral of some $f \in L^1(\mathbb{R}^{n-1}, \sigma)$, where $d\sigma = (1 + \parallel y \parallel^n)^{-1} d^{n-1}y$, and there are constants $A = A(h) > 0$ and $\alpha = \alpha(h) > 0$ such that for all $x \in \mathbb{R}^n_+$ and all $\lambda > 0$*

$$\mu_x\{y \in \mathbb{R}^{n-1} : |f(y) - h(x)| > \lambda\} \leqq A e^{-\alpha\lambda},$$

where μ_x denotes harmonic measure with respect to $x \in \mathbb{R}^n_+$;

(vi) *for some constant $\beta = \beta(h) > 0$ the subharmonic function $e^{\beta|h|}$ admits a harmonic majorant and*

$$\sup_{x \in \mathbb{R}^n_+} M\, e^{\beta|h - h(x)|}(x) < \infty;$$

(vii) *there are constants $B = B(h) > 0$ and $\beta = \beta(h) > 0$ such that the subharmonic function $e^{\beta h}$ admits a harmonic majorant, satisfying*

$$M\, e^{\beta h} \leqq B\, e^{\beta h}; \tag{3.2}$$

(viii) *there is a positive harmonic function u on $\mathbb{R}^n_+$ and a constant $\delta = \delta(h) > 0$ such that the subharmonic function $h - \delta \log u$ is bounded on $\mathbb{R}^n_+$.*

Proof.

(i) $\Rightarrow$ (v): Combine Theorems 1.2 and 2.1.

(v) $\Rightarrow$ (iii): For arbitrary $x \in \mathbb{R}^n_+$,

$$M|h - h(x)|^p(x) = \int_{\mathbb{R}^{n-1}} |f(y) - h(x)|^p P_x(y) d^{n-1}y$$

$$= p \int_0^\infty \mu_x\{y \in \mathbb{R}^{n-1} : |f(y) - h(x)| > \lambda\}\lambda^{p-1}d\lambda$$

$$\leqq p \int_0^\infty A\, e^{-\alpha\lambda}\lambda^{p-1}d\lambda = Ap\Gamma(p)/\alpha^p.$$

Hence

$$\parallel h \parallel_{*,p} \leqq \frac{(Ap\Gamma(p))^{1/p}}{\alpha}. \tag{3.3}$$

(iii) $\Rightarrow$ (ii): trivial

(ii) $\Rightarrow$ (i): follows from $M|h - h(x)| \leq M|h - h(x)|^p + 1$

(iii) $\Rightarrow$ (iv): Note that

$$M|h - h(x)|^2(x) = Mh^2(x) - h^2(x), \qquad (3.4)$$

for all $x \in \mathbb{R}^n_+$. Hence, by hypothesis, the superharmonic function $v = Mh^2 - h^2$ is bounded on $\mathbb{R}^n_+$. Since it obviously is a potential (having only negative harmonic minorants) it admits the representation

$$v(x) = \text{const} \cdot \int_{\mathbb{R}^n_+} G(x, y)(\Delta v)(y)d^n y,$$

and hence, on account of $\Delta v = -\Delta(h^2) = -2 \parallel \text{grad} \, h \parallel^2$, we are done.

(iv) $\Rightarrow$ (ii): Clearly, by (3.4), the condition (3.1) is satisfied for $p = 2$.

(v) $\Rightarrow$ (vi): For arbitrary $0 < \beta < \alpha$,

$$M \, e^{\beta|h - h(x)|}(x) = \int_{\mathbb{R}^{n-1}} e^{\beta|f(y) - h(x)|} P_x(y)d^{n-1}y$$

$$= \beta \int_0^\infty \mu_x\{y \in \mathbb{R}^{n-1} : |f(y) - h(x)| > \lambda\} \cdot e^{\beta\lambda}d\lambda$$

$$\leqq \beta \int_0^\infty A \, e^{-\alpha\lambda}e^{\beta\lambda}d\lambda = \frac{A\beta}{\alpha - \beta}.$$

(vi) $\Rightarrow$ (vii): Set

$$B = \sup_{x \in \mathbb{R}^n_+} M \, e^{\beta|h - h(x)|}(x).$$

Then

$$B = M \, e^{\beta(h - h(x))}(x) = e^{-\beta h(x)}M \, e^{\beta h}(x)$$

and thus (3.2) holds.

(vii) $\Rightarrow$ (i): From the elementary inequality $x^+ \leqq e^x$, valid for all $x \in \mathbb{R}$, we obtain

$$\beta(h - h(x))^+ \leqq e^{\beta(h - h(x))} \leqq e^{-\beta h(x)}M \, e^{\beta h}.$$

Consequently

$$\beta M(h - h(x))^+(x) \leqq e^{-\beta h(x)}(M \, e^{\beta h})(x) \leqq B,$$

for all $x \in I\!\!R^n_+$. In view of the fact that $M(h - h(x))^+(x) = \frac{1}{2}M|h - h(x)|(x)$ (see [10], (1.7)) this implies that

$$\| h \|_* \leqq 2\frac{B}{\beta}. \tag{3.5}$$

(vii) $\Rightarrow$ (viii): Put $u = M\, e^{\beta h}$. Then, by hypothesis, $u/B \leq e^{\beta h} \ (\leq u)$, and hence $h - \frac{1}{\beta} \log u$ is bounded.

(viii) $\Rightarrow$ (vii): From

$$\delta \log u - c \leqq h \leqq \delta \log u + c$$

we conclude that

$$M\, e^{h/\delta} \leqq e^{c/\delta} u \leqq e^{2c/\delta} e^{h/\delta},$$

finishing the proof of Theorem 3.1. $\square$

Remark. The equivalence of the statements (i) – (iii) and (vi) – (viii) holds true on any domain in $I\!\!R^n$, in fact on an arbitrary Brelot space (X, H) with $1 \in H(X)$. The proof depends on a result of T. Lyons [13] shown by making use of the martingale version of the John-Nirenberg theorem. It would be desirable, however, to find a non-probabilistic proof of this result.

All norms $\| h \|_{*,p}$ introduced in (3.1), are *equivalent,* as we are going to show next.

First recall the *Hardy space* h^p $(1 \leqq p < \infty)$ of harmonic functions defined on some domain in $\Omega \in I\!\!R^n$,

$$h^p(\Omega) = \{h \in H(\Omega) : |h|^p \quad \text{admits a harmonic majorant}\}.$$

Endowed with the norm

$$\| h \|_p = _{x_0}\| h \|_p = (M|h|^p(x_0))^{1/p},$$

defined relative to some (fixed) point $x_0 \in \Omega$, it represents a well-known Banach space. Note that

$$\| h \|_{*,p} = \sup_{x_0 \in \Omega} {}_{x_0}\| h - h(x_0) \|_p .$$

What we are going to need in what follows is the inequality

$$\| h \|_1 \leqq \| h \|_p \tag{3.6}$$

valid for all $h \in h^p$ ($1 \leqq p < \infty$). Its proof depends on the estimate

$$px \leqq (p-1)u + x^p u^{1-p}$$

which holds for all $x, u \in \mathbb{R}_+$. From

$$pM|h| \leqq (p-1)u + u^{1-p}M|h|^p$$

we conclude that

$$p \parallel h \parallel_1 \leqq (p-1)u + u^{1-p} \parallel h \parallel_p^p$$

and hence the choice of $u = \parallel h \parallel_p$ yields (3.6).

Applied to BMOH ($\mathbb{R}_+^n$) the inequality (3.6) yields

$$\parallel h \parallel_* \leqq \parallel h \parallel_{*,p} . \tag{3.7}$$

On the other hand, (3.3), (2.1), and (1.8) show that

$$\parallel h \parallel_{*,p} \leqq (Ap\Gamma(p))^{1/p} \frac{\parallel f \parallel_*}{b} \leqq (Ap\Gamma(p))^{1/p} \frac{c}{b} \parallel h \parallel_* . \tag{3.8}$$

Hence all norms $\parallel h \parallel_{*,p}$ ($1 \leqq p < \infty$) are shown to be equivalent.

Observe that as a consequence the square root of the supremum norm of the Green potential in statement (iv) of Theorem 3.1 also defines a norm equivalent to $\parallel h \parallel_*$.

In view of Theorem 3.1 it is easy to find examples of unbounded functions of class BMOH, provided one makes use of the following observation:
For any $h, \tilde{h} \in h^2(\Omega)$ (Ω a domain in $\mathbb{R}^n$) with the property that $h^2 - \tilde{h}^2$ is subharmonic we have

$$M\tilde{h}^2 - \tilde{h}^2 \leq Mh^2 - h^2. \tag{3.9}$$

Equality holds if and only if $h^2 - \tilde{h}^2$ is harmonic on Ω.

Now by (3.4) and Theorem 3.1, $h \in$ BMOH if and only if $Mh^2 - h^2$ is bounded. Hence if $h \in$ BMOH (e.g. if h is bounded) any function $\tilde{h} \in h^2$ with the property that $h^2 - \tilde{h}^2$ is subharmonic lies in BMOH.

To verify (3.9) just consider the inequality $\tilde{h}^2 \leq Mh^2 + \tilde{h}^2 - h^2$. Since the left side is subharmonic, the right one however superharmonic, it follows that $M\tilde{h}^2 \leqq Mh^2 + \tilde{h}^2 - h^2$, i.e. (3.9) holds.

In case $n = 2$ we get the well known result that the harmonic conjugate $\tilde{h}$ of any $h \in$ BMOH is again in this class. Furthermore, noting that $\parallel \tilde{h} \parallel_{*,2} = \parallel h \parallel_{*,2}$, the relations (3.7) and (3.8) show that there is a constant $c > 0$ depending only on n, such that $c^{-1} \parallel h \parallel_* \leqq \parallel \tilde{h} \parallel_* \leqq c \parallel h \parallel_*$.

§4 Lipschitz continuity of BMOH with respect to the hyperbolic metric.

Let us first remark that the equivalence of statements (i) and (viii) in Theorem 3.1 may be specified as follows:

Theorem 4.1. *A harmonic function h on $I\!\!R^n_+$ is in BMOH if and only if there is a positive harmonic function v on $I\!\!R^n_+$ and a constant $\gamma = \gamma(h) > 0$ such that*

$$\gamma(\log v - 1) \leqq h \leqq \gamma(\log v + 1) \tag{4.1}$$

holds on $I\!\!R^n_+$.

In the light of Theorem 3.1 all we have to prove is a lemma of the following type

Lemma 4.2. *Let h be harmonic on a domain $\Omega \subset I\!\!R^n$. Assume that there is a positive harmonic function u on Ω and constants $\delta, c > 0$ such that*

$$\delta \log u - c \leqq h \leqq \delta \log u + c.$$

Then there is a positive harmonic function v on Ω such that (4.1) holds, provided we set $\gamma = \max(\delta, c)$.

Proof. Obviously, if $c \leqq \delta$ we may take $v = u$. Hence assume that $c > \delta$ and put $\eta = \delta/c$. Since $0 < \eta < 1$ the function u^η is superharmonic and hence admits a greatest harmonic minorant denoted by $m\, u^\eta$. From $\log u = (h-c)/\delta$ we conclude that the superharmonic function $\log u$ also admits a greatest harmonic minorant. Denoting it by $m \log u$, we have

$$h - c \leqq \delta(m \log u). \tag{4.2}$$

From $e^{\eta(m \log u)} \leqq u^\eta$, and the subharmonicity of the function on the left hand side, we obtain the inequality $e^{\eta(m \log u)} \leqq m\, u^\eta$. Setting $v = m\, u^\eta$ we thus showed that

$$\delta(m \log u) \leqq c \log v \leqq \delta \log u.$$

Combining this result with (4.2) we infer from our hypothesis that

$$c(\log v - 1) \leqq \delta \log u - c \leqq h \leqq \delta(m \log u) + c \leqq c(\log v + 1),$$

completing the proof of Lemma 4.2. $\square$

Note that this proof actually works on an arbitrary harmonic space and in fact with u superharmonic instead of u harmonic.

Theorem 4.1 allows the introduction of a further norm on BMOH, namely

$$\| h \|_\circ = \inf\{\gamma > 0 : \exists v \in H^+ \quad \text{such that (4.1) holds}\}, \qquad (4.3)$$

where H^+ denotes the set of positive harmonic functions on $\mathbb{R}^n_+$.

Theorem 4.3. *The norms $\| h \|_\circ$ and $\| h \|_*$ are equivalent, i.e. there is a constant $D > 0$, depending only on the dimension n, such that*

$$\frac{1}{D} \| h \|_* \leqq \| h \|_\circ \leqq D \| h \|_*,$$

for all $h \in$ BMOH $(\mathbb{R}^n_+)$.

Proof. Let $h \in$ BMOH $(\mathbb{R}^n_+)$. The proof of Theorem 3.1, more precisely of the implications (i) $\Rightarrow$ (v) $\Rightarrow$ (vi) $\Rightarrow$ (vii), shows that there are constants a and b, depending only on n, such that

$$M\, e^{(\alpha/2)h} \leqq a\, e^{(\alpha/2)h},$$

where $\alpha = b/\| f \|_*$ (f denoting the boundary function of h). Hence, setting $u = M\, e^{(\alpha/2)h}$, we obtain

$$\frac{2}{\alpha} \log \frac{u}{\sqrt{a}} - \frac{1}{\alpha} \log a \leq h \leq \frac{2}{\alpha} \log \frac{u}{\sqrt{a}} + \frac{1}{\alpha} \log a.$$

An application of Lemma 4.2 then yields a $v \in H^+$ such that (4.1) holds, provided we set $\gamma = \max(2/\alpha, (\log a)/\alpha)$.

Consequently

$$\gamma = \frac{\max(2, \log a)}{b} \| f \|_* = \| h \|_\circ,$$

and thus invoking (1.8) yields $\| h \|_\circ \leqq (c/b) \max(2, \log a) \| h \|_*$. On the other hand, if (4.1) holds, the proof of the implication (viii) $\Rightarrow$ (vii) shows that

$$M\, e^{h/\gamma} \leqq e^2 e^{h/\gamma}.$$

It then follows from (3.5) that $\| h \|_* \leqq 2e^2 \gamma$ and hence $\| h \|_* \leqq 2e^2 \| h \|_\circ$, completing the proof of Theorem 4.3. $\square$

Theorem 4.1 shows that there is a strong connection of BMOH with the so-called *Harnack distance* defined for an arbitrary domain Ω in $I\!\!R^n$ as follows:

$$\left.\begin{aligned} d(x,y) = d_\Omega(x,y) &= \log \inf\{a|a^{-1} \leqq h(x)/h(y) \leqq a : \forall h \in H^+(\Omega)\} \\ &= \sup\{|\log h(x) - \log h(y)| : h \in H^+(\Omega)\} \end{aligned}\right\} \quad (4.4)$$

In [12] this distance has been determined explicitly in the case of $\Omega = I\!\!R^n_+$, but we shall not need this result here. It will be sufficient to know that in $I\!\!R^n_+$ ([12], (5.4))

$$d \leqq n\, d_{\mathrm{hyp}}, \qquad\qquad (4.5)$$

where d_{hyp} denotes the *hyperbolic distance*, defined by the Poincaré metric

$$ds = \frac{\|\, dx\, \|}{x_n}.$$

What we shall need, however, is the following

Lemma 4.4. *Let Ω be a domain in $I\!\!R^n$ and $d = d_\Omega$ its Harnack distance. Then for any bounded harmonic function h on Ω we have*

$$|h(x) - h(y)| \leqq (e^{d(x,y)} - 1) \sup_{\xi,\eta\in\Omega} |h(\xi) - h(\eta)|,$$

for all $x, y \in \Omega$.

Proof. Set $c = \sup\limits_{\xi,\eta\in\Omega} |h(\xi) - h(\eta)|$ and consider the harmonic function $u(x) = h(x) - h(y) + c$, for fixed $y \in \Omega$. Since $u = 0$, (4.4) implies that

$$u(x) \leqq e^{d(x,y)} u(y)$$

and hence

$$h(x) - h(y) = u(x) - c \leqq (e^{d(x,y)} - 1)c.$$

Interchanging x and y yields the desired result. $\square$

From (4.1) and (4.4) we see that for every $h \in \mathrm{BMOH}\,(I\!\!R^n_+)$

$$\left.\begin{aligned} |h(x) - h(y)| &\leqq \gamma(|\log v(x) - \log v(y)| + 2) \\ &\leqq \gamma(d(x,y) + 2), \end{aligned}\right\} \qquad (4.6)$$

for all $x, y \in I\!\!R^n_+$. Hence, by (4.3) and (4.5)

$$|h(x) - h(y)| \leq \| h \|_0 \, (d(x, y) + 2) \leq \| h \|_0 \, (n \, d_{\mathrm{hyp}}(x, y) + 2), \qquad (4.7)$$

for all $x, y \in I\!\!R^n_+$. In order to incorporate the number 2 in the parenthesis into a multiplicative constant we first note that for all $x, y \in I\!\!R^n_+$ with $d_{\mathrm{hyp}}(x, y) = 1/n$ we have

$$|h(x) - h(y)| \leq 3n \, \| h \|_\circ \, d_{\mathrm{hyp}}(x, y).$$

Next we consider the non-euclidean ball

$$B_{\mathrm{hyp}} = B_{\mathrm{hyp}} \left(y, \frac{2}{n} \right) = \left\{ x \in I\!\!R^n_+ : d_{\mathrm{hyp}}(x, y) < \frac{2}{n} \right\}.$$

On B_{hyp} we have, on account of (4.7),

$$|h(x) - h(y)| \leq \| h \|_\circ \, (n \, d_{\mathrm{hyp}}(x, y) + 2) \leq 4 \, \| h \|_\circ \, .$$

From Lemma 4.4, applied to the ball B_{hyp} we then conclude that for all $x, y \in B_{\mathrm{hyp}}$:

$$|h(x) - h(y)| \leq 4 \, \| h \|_\circ \, (e^{d(x,y)} - 1) \leq 4 \, \| h \|_\circ \, (e^{\rho(x,y)} - 1),$$

where $\rho(x, y)$ denotes the hyperbolic distance of x and y, defined relative to the ball B_{hyp}, multiplied by n. Mapping $I\!\!R^n_+$ conformally onto the unit ball B_n in $I\!\!R^n$ by a Möbius transformation Φ with $\Phi(y) = 0$, the balls $B_{\mathrm{hyp}}(y, \frac{1}{n})$ and $B_{\mathrm{hyp}}(y, \frac{2}{n})$ will be mapped onto concentric balls in B_n. From this remark we conclude that there is a constant $K > 0$, independent of y, such that for all $x \in B_{\mathrm{hyp}}(y, \frac{1}{n})$

$$e^{\rho(x,y)} - 1 \leq K \, d_{\mathrm{hyp}}(x, y).$$

Consequently, for all $x \in I\!\!R^n_+$ with $d_{\mathrm{hyp}}(x, y) < 1/n$

$$|h(x) - h(y)| \leq 4K \, \| h \|_\circ \, d_{\mathrm{hyp}}(x, y).$$

We thus proved that

$$|h(x) - h(y)| \leq \mathrm{const} \, \| h \|_\circ \, d_{\mathrm{hyp}}(x, y)$$

for all $x, y \in I\!\!R^n_+$, with a constant depending only on n. Combining this result with Theorem 4.3 we obtain

Theorem 4.5. *Let $h \in \text{BMOH}\,(I\!\!R^n_+)$. There exists a constant $C > 0$, depending only on the dimension n, such that*

$$|h(x) - h(y)| \leqq C \parallel h \parallel_* d_{\text{hyp}}(x, y),$$

for all $x, y \in I\!\!R^n_+$.

Remark. The same result holds true in case $I\!\!R^n_+$ is replaced by the unit ball B_n, as the proof of Theorem 4.5 shows. It thus extends Theorem 3.1 in [11].

Let us also mention

Corollary 4.6. *Let $h \in \text{BMOH}\,(I\!\!R^n_+)$. There is a constant $C > 0$ depending only on n, such that*

$$\parallel \text{grad}\, h \parallel (x) \leqq \frac{C}{x_n} \parallel h \parallel_*,$$

for all $x = (x_1, \ldots, x_n) \in I\!\!R^n_+$.

Introducing, with respect to an arbitrary doamin $\Omega \subset I\!\!R^n$, the Bloch space

$$\text{BLOCH}\,(\Omega) = \{h \in H(\Omega) : \sup_{x \in \Omega} \text{dist}\,(x, \partial\Omega) \parallel \text{grad}\, h \parallel (x) < \infty\},$$

where $\text{dist}\,(x, \partial\Omega)$ denotes the euclidean distance from x to the boundary $\partial\Omega$, it is not difficult to verify that $\text{BLOCH}\,(\Omega)$ contains all bounded harmonic functions on Ω. However, as we see from Corollary 4.6, in case $\Omega = I\!\!R^n_+$ (resp. B_n) we even have

$$\text{BMOH}\,(I\!\!R^n_+) \subset \text{BLOCH}\,(I\!\!R^n_+).$$

This extends a result, known in $I\!\!R^2_+$, to $I\!\!R^n_+$ $(n > 2)$.

§5 On BMO-domains and the torsion function

The purpose of this section is to extend to $I\!\!R^n$ $(n = 2)$ the concept of a BMO-domain, introduced in the complex plane $C\!\!\!\!C$ by A. Baernstein [1].

Recall the definition of a BMO-domain $\Omega \subset C\!\!\!\!C$. It is based on the class BMOA of analytic functions of bounded mean oscillation, defined on the unit disc B_2 as follows:

$$\text{BMOA} = \{f : f \quad \text{analytic on} \quad B_2, \ \text{Re}\, f \in \text{BMOH}\,(B_2)\}, \qquad (5.1)$$

where $\operatorname{Re} f$ denotes the real part of f. Note that according to Theorem 3.1 and (3.4)

$$f \in \mathrm{BMOA} \iff Mh^2 - h^2 \ (h = \operatorname{Re} f) \quad \text{is bounded on} \quad B_2. \tag{5.2}$$

Denoting the imaginary part $\operatorname{Im} f$ by $\tilde{h}$ we conclude from (3.9) that

$$Mh^2 - h^2 = M\tilde{h}^2 - \tilde{h}^2 = \frac{1}{2}(M|f|^2 - |f|^2).$$

Consequently

$$f \in \mathrm{BMOA} \iff \operatorname{Im} f \in \mathrm{BMOH}\,(B_2)$$
$$\iff M|f|^2 - |f|^2 \quad \text{is bounded on} \quad B_2 \tag{5.3}$$
$$\iff \sup_{z \in B_2} M|f - f(z)|^2(z) < \infty.$$

By definition $\Omega \subset \boldsymbol{C}$ is called a *BMO-domain*, provided that every analytic function f on B_2 whose image set $f(B_2)$ is contained in Ω belongs to the class BMOA.

Obviously, every bounded domain is a BMO-domain, but there are also unbounded ones.

Of course this definition of a BMO-domain can not be extended to higher dimensions. However

Lemma 5.1. *A domain $\Omega \subset \boldsymbol{C}$ is a* BMO-*domain if and only if the following conditions hold:*

a) *The subharmonic function $w \to |w|^2$, restricted to Ω, admits a harmonic majorant and*

b) *the least harmonic majorant, $M_\Omega|w|^2$, satisfies the condition:*

$$M_\Omega|w|^2 - |w|^2 \quad \text{is bounded on} \quad \Omega.$$

Proof. (Sufficiency). Let $f : B_2 \to \boldsymbol{C}$ be an analytic function with $f(B_2) \subset \Omega$. Put $H = M_\Omega|w|^2$ and observe that $|f|^2 \leq H \circ f$ implies $M|f|^2 \leq H \circ f$, where M signifies taking the least harmonic majorant on B_2. By hypothesis there is a constant $C > 0$ such that $H(w) - |w|^2 \leq C$, for all $w \in \Omega$. Hence $M|f|^2 - |f|^2 \leq H \circ f - |f|^2 \leq C$, and thus, by (5.3), $f \in \mathrm{BMOA}$.

(Necessity). Let $\Omega \subset \boldsymbol{C}$ be a BMO-domain. Considering Ω as a planar Riemannian surface it is of hyperbolic type, i.e. its universal covering surface is the unit disc B_2. Let $f : B_2 \to \Omega$ denote the universal covering map and Γ

the group of deck transformations. Since by hypothesis Ω is a BMO-domain, we have $f \in \mathrm{BMOA}$. Thus by (5.3) the function $M|f|^2 - |f|^2$ is bounded on B_2. Now all we have to show is that with f invariant under Γ so is $M|f|^2$. Indeed, if $M|f|^2 = H \circ f$, for some H harmonic on Ω, then $H(w) = |w|^2$, for all $w \in \Omega$, and hence $(M_\Omega |w|^2)(w) - |w|^2 \leq H(w) - |w|^2 \leq \mathrm{const.}$, for all $w \in \Omega$. $\square$

The invariance of $M|f|^2$ under Ω follows from

Lemma 5.2. *Let U be open in $\mathbb{C}$, resp. on a Riemannian surface, and Φ a biholomorphic mapping of U onto itself. Further, let u be subharmonic on U and assume that u admits a harmonic majorant. Then $u \circ \Phi = u$ implies that $(Mu) \circ \Phi = Mu$, where Mu denotes the least harmonic majorant of u on U.*

Proof. Set $H = (Mu) \circ \Phi$. Then $H = u \circ \Phi = u$ and hence $H = Mu$. Consider $\tilde{H} = (Mu) \circ \Phi^{-1}$. Then $\tilde{H} \leq H \circ \Phi^{-1} = Mu$. On the other hand, $\tilde{H} = u \circ \Phi^{-1} = u$ implies that $\tilde{H} = Mu$. Consequently $\tilde{H} = Mu$ and thus $Mu = H$. $\square$

Identifying $\mathbb{C}$ with $\mathbb{R}^2$ and denoting the ith coordinate function $(w_1, w_2) \to w_i$ by φ_i $(i = 1, 2)$, (3.9) shows that

$$\left. \begin{aligned} M_\Omega \varphi_1^2 - \varphi_1^2 &= M_\Omega \varphi_2^2 - \varphi_2^2 \\ &= \frac{1}{2}(M_\Omega |w|^2 - |w|^2). \end{aligned} \right\} \tag{5.4}$$

Consequently, by Lemma 5.1,

$$\Omega \quad \text{is a BMO} - \text{domain} \iff \text{for} \quad i = 1, 2, : \varphi_i \in \mathrm{BMOH}\,(\Omega). \tag{5.5}$$

This is the characterization which we are going to extend to higher dimensions.

Definition. A domain $\Omega \subset \mathbb{R}^n$ $(n = 2)$ is called a *BMO-domain* if each coordinate function $\varphi_i : (x_1, \ldots, x_n) \to x_i$ $(i = 1, \ldots, n)$, restricted to Ω, is in class $\mathrm{BMOH}\,(\Omega)$.

Obviously, bounded domains are BMO-domains. Furthermore, subdomains of BMO-domains are BMO-domains. The union of two BMO-domains, however, is generally not a BMO-domain.

Observe that conditions a) and b) in Lemma 5.1 also extend to higher dimensions. Indeed, if $\Omega \subset \mathbb{R}^n$ $(n = 2)$ we may consider the following conditions:

a) The subharmonic function $x \to \|x\|^2$, restricted to Ω, admits a harmonic majorant, and,

b) denoting the least harmonic majorant by $M_\Omega \|x\|^2$, the function

$$M_\Omega \|x\|^2 - \|x\|^2 \quad \text{is bounded on} \quad \Omega. \tag{5.6}$$

Noting that for any (fixed) point $x_0 \in \Omega$

$$M_\Omega \|x - x_0\|^2 = M_\Omega \|x\|^2 - 2(x, x_0) + \|x_0\|^2,$$

we find that

$$M_\Omega \|x - x_0\|^2 (x_0) = M_\Omega \|x\|^2 (x_0) - \|x_0\|^2. \tag{5.7}$$

Hence (5.5) holds if and only if

$$\sup_{x_0 \in \Omega} M_\Omega \|x - x_0\|^2 (x_0) < \infty. \tag{5.8}$$

Clearly, for any domain $\Omega \subset \mathbb{R}^n$ with the property that its topological boundary agrees with its Martin boundary we have

$$M_\Omega \|x - x_0\|^2 (x_0) = \int_{\partial \Omega} r^2 d\mu_{x_0}, \tag{5.9}$$

where $r = \|x - x_0\|$ and μ_{x_0} denotes the harmonic measure with respect to $x_0 \in \Omega$.

Consequently the function $x_0 \to M_\Omega \|x - x_0\|^2 (x_0)$ is, except for a multiplicative constant, precisely what J. Hersch [8] calls the *torsion function* of Ω. Conditions a) and b) thus simply say that on Ω the torsion function exists and is bounded (resp., in the notation of J. Hersch [8], that $t_{\max} < \infty$). This suggests the following

Definition. A domain $\Omega \subset \mathbb{R}^n$ $(n = 2)$ is called a *BT-domain (bounded torsion)* if the above conditions a) and b) hold.

Note that on account of the isoperimetric inequality proved by J. Hersch in ([8], formula (15)) for $n = 3$ and by M. Sakai ([15], Theorem 1.4) for arbitrary $n \in \mathbb{N}$, we have the following result: *Every domain $\Omega \subset \mathbb{R}^n$ of finite volume is a BT-domain.* But there are also BT-domains of infinite volume (e.g. cylinders) as the following characterization shows. It is based on a fundamental result (see the equivalence of (i) and (ix) below) proved in $\mathbb{C}$ by Hayman-Pommerenke [7] (and independently by Stegenga [16]) and extended to $\mathbb{R}^n$ recently by Banuelos-Øksendal [2].

Theorem 5.3. *Let Ω be a domain in $\mathbb{R}^n$ ($n = 2$). The following statements are equivalent:*

(i) Ω *is* BMO*-domain.*

(ii) Ω *is a* BT*-domain (bounded torsion).*

(iii) *for some $i \in \{1, \ldots, n\}$ the coordinate function $\varphi_i : (x_1, \ldots, x_n) \to x_i$, restricted to Ω, is in class* BMOH (Ω).

(iv) *there is a superharmonic function s on Ω and a constant $C > 0$ such that*

$$\| x \|^2 \leqq s(x) \leqq \| x \|^2 + C,$$

for all $x \in \Omega$.

(v)   *there is a bounded solution $u \in C^2(\Omega)$ of the equation $\Delta u + 1 = 0$.*

(vi) *there is a bounded lower semicontinuous function $v : \Omega \to \mathbb{R} \cup \{+\infty\}$ such that $\Delta v + 1 \leqq 0$ (in the sense of distributions)*

(vii) *the Green potential*

$$\int_\Omega G_\Omega(\cdot, y) d^n y$$

is bounded on Ω (G_Ω = Green's function for Ω)

(viii) *Let B_t denote Brownian motion on $\mathbb{R}^n$ and τ_Ω the first exit time, i.e.*

$$\tau_\Omega(\omega) = \inf\{t > 0 : B_t(\omega) \notin \Omega\}.$$

Then the function $x \to E^x \tau_\Omega$ is bounded on Ω.

(ix) *There are constants $R, \delta > 0$ such that*

$$\mathrm{cap}\,[B(x, R) \setminus \Omega] = \delta, \tag{5.10}$$

for all $x \in \Omega$. Hereby cap *denotes Newtonian capacity with respect to the ball $B(x, 2R)$, if $n = 3$, whereas for $n = 2$ it designates logarithmic capacity.*

(x) Ω *is quasi-regular (in the sense of T. Sturm [17]).*

Remark. According to T. Sturm [17] a domain Ω is called *quasi-regular* if for all bounded, Borel-measurable functions g on Ω the Dirichlet problem for the Poisson equation $\Delta u = g$ has a unique, bounded solution. More precisely, given a bounded continuous function f on $\partial\Omega$ there is a unique bounded continuous solution u of $\Delta u = g$ such that $\lim_{x \to y} u(x) = f(y)$ at all regular boundary points $y \in \partial\Omega$ (regular with respect to the Laplace equation $\Delta u = 0$).

Proof. (ii) $\Rightarrow$ (i): Analogously to (3.4) and (5.4) one verifies that

$$M(\varphi_i - \varphi_i(x))^2(x) = M\varphi_i^2(x) - \varphi_i^2(x) = \frac{1}{n}(M \parallel x \parallel^2 (x)- \parallel x \parallel^2) \quad (5.11)$$

for all $x \in \Omega$. Hence (i) follows from

$$M|\varphi_i - \varphi_i(x)| \leq 1 + M(\varphi_i - \varphi_i(x))^2 \quad (x \in \Omega),$$

for $i = 1, \ldots, n$.

(i) $\Rightarrow$ (iii): trivial.

(iii) $\Rightarrow$ (ii): According to the remark following the proof of Theorem 3.1 $\varphi_i \in \mathrm{BMOH}\,(\Omega)$ for some $i \in \{1, \ldots, n\}$ implies that

$$\sup_{x\in\Omega} M(\varphi_i - \varphi_i(x))^2 < \infty$$

and hence (ii) follows from (5.11).

(ii) $\Rightarrow$ (iv): Choose $s = M \parallel x \parallel^2$.

(iv) $\Rightarrow$ (ii): From $\parallel x \parallel^2 \leq s(x)$, for all $x \in \Omega$, we conclude that $M \parallel x \parallel^2 \leq s$. Hence $M \parallel x \parallel^2 - \parallel x \parallel^2 \leq s- \parallel x \parallel^2 \leq C$.

(ii) $\Rightarrow$ (v): Set $u = \frac{1}{2n}(M \parallel x \parallel^2 - \parallel x \parallel^2)$. Then $u \in C^2(\Omega)$ is a bounded solution of the equation $\Delta u + 1 = 0$.

(v) $\Rightarrow$ (ii): Let u be a bounded solution of $\Delta u + 1 = 0$. W.l.o.g. we may assume that $u = 0$. Put $h(x) = 2n\,u(x)+ \parallel x \parallel^2$. Then $\Delta h = 0$, i.e. h is harmonic, and

$$\parallel x \parallel^2 \leq h(x) \leq \parallel x \parallel^2 +C,$$

for all $x \in \Omega$ and some constant $C > 0$. Consequently $M \parallel x \parallel^2 \leq h$ and thus

$$(M \parallel x \parallel^2)(x)- \parallel x \parallel^2 \leq h(x)- \parallel x \parallel^2 \leq C,$$

for all $x \in \Omega$.

(iv) $\Rightarrow$ (vi): Set $v(x) = \frac{1}{2n}(s(x)- \parallel x \parallel^2)$. Then v is bounded and $\Delta v + 1 = (1/2n)\Delta s \leq 0$.

(vi) $\Rightarrow$ (iv): Put $s(x) = 2nv(x)+ \parallel x \parallel^2$. W.l.o.g. $0 \leq v \leq$ const.. Hence $\parallel x \parallel^2 \leq s(x) \leq \parallel x \parallel^2 +$ const. and $\Delta s = 2n(\Delta v + 1) \leq 0$, showing that s is superharmonic.

(ii) $\Rightarrow$ (vii): Set $w = M \parallel x \parallel^2 - \parallel x \parallel^2$. As remarked earlier, w is a potential and hence $\Delta w = -2n$ implies that

$$w(x) = \mathrm{const} \int_{\Omega} G_\Omega(x, y)\, d^n y, \quad (5.12)$$

showing that the integral is bounded.

(vii) $\Rightarrow$ (viii): Use the formula of Hunt (see e.g. [4]):

$$E^x \tau_\Omega = \int_\Omega G_\Omega(x, y) d^n y. \tag{5.13}$$

(viii) $\Rightarrow$ (ii): As shown by Burckholder [3], $M \parallel x \parallel^2$ exists. Set

$$w = M \parallel x \parallel^2 - \parallel x \parallel^2 .$$

Then the representation (5.12) holds and hence (5.13) shows that w is bounded.

(ix) $\Leftrightarrow$ (i) (for $n = 2$): See Hayman-Pommerenke [7], resp. Stegenga [16].

(ix) $\Leftrightarrow$ (viii) (for $n > 2$): See Banuelos-Øksendal ([2], Theorem 2.1).

(vii) $\Leftrightarrow$ (x): See Sturm ([17], Proposition 3.1).

Thus the proof of Theorem 5.3 is complete. $\square$

Generalization. Replacing the function $x \to \parallel x \parallel^2 \colon I\!R^n \to I\!R$, by an arbitrary (fixed) subharmonic function Ψ on a harmonic space (X, H), the concept of a BMO-domain may be generalized by looking at those domains $\Omega \subset X$ with the property that

a) the restriction of Ψ to Ω admits a harmonic majorant, and

b) the least harmonic majorant $M_\Omega \Psi$ satisfies the condition $M_\Omega \Psi - \Psi \leqq$ const.

Note however that any two subharmonic functions Ψ and $\tilde{\Psi}$ on X, with $\Psi - \tilde{\Psi}$ harmonic, define the same class of domains Ω. Indeed, since $\Psi - \tilde{\Psi}$ is harmonic, we have

$$M_\Omega \Psi - \Psi = M_\Omega \tilde{\Psi} - \tilde{\Psi},$$

in consequence of (3.9).

References

[1] Baernstein, A.II: Analytic functions of bounded mean oscillation. Aspects of contemporary complex analysis (Durham, 1979), 3–36, Academic Press, London 1980.

[2] Bañuelos, R. and Øksendal, B.: Exit times for elliptic diffusions and BMO. Proc. Edinburgh Math. Soc. 30 (1987), 273–288.

[3] Burkholder, D.L.: Exit times of Brownian motion, harmonic majorization, and Hardy spaces. Advances in Math. 26 (1977), 182–205.

[4] Durrett, R.: Brownian Motion and Martingales in Analysis. Wadsworth, 1984.

[5] Fefferman, C. and Stein, E.M.: H^p-spaces of several variables, Acta Math. 129 (1972), 137–193.

[6] Garnett, J.B.: Bounded analytic functions. Academic Press, New York, 1981.

[7] Hayman, W.K. and Pommerenke, Ch.: On analytic functions of bounded mean oscillation. Bull. London Math. Soc. 10 (1978), 219–224.

[8] Hersch, J.: On the torsion function, Green's function and conformal radius: an isoperimetric inequality of Pólya and Szegö, some extensions and applications. J. Analyse Math. 36 (1979), 102–117.

[9] John, F. and Nirenberg, L.: On functions of bounded mean oscillation. Comm. Pure Appl. Math. 14 (1961), 415–426.

[10] Leutwiler, H.: Harmonic functions of bounded mean oscillation. Math. Ann. 244 (1979), 167–183.

[11] Leutwiler, H.: Harmonic functions of bounded mean oscillation and a generalization to vector lattices of continuous functions. Potential theory (Copenhagen, 1979), 194–208, Lecture Notes in Math., 787, Springer, Berlin 1980.

[12] Leutwiler, H.: On a distance invariant under Möbius transformations in $I\!R^n$. Ann. Acad. Sci. Fenn. Ser. A I Math. 12 (1987), 3–17.

[13] Lyons, T.: A definition of BMO_p for an abstract harmonic space and a John-Nirenberg theorem. Bull. London Math. Soc. 12 (1980), 127–129.

[14] Reimann, H.M. and Rychener, T.: Funktionen beschränkter mittlerer Oszillation. Lecture Notes in Math., 487, Springer, Berlin 1975.

[15] Sakai, M.: Isoperimetric inequalities for the least harmonic majorant of $|x|^p$. Trans. Am. Math. Soc. 299 (1987), 431–472.

[16] Stegenga, D.A.: A geometric condition which implies BMOA. Harmonic analysis in Euclidean spaces (Williamstown, Mass. 1978), Part 1, 427–430, Proc. Sympos. Pure Math. 35:1, Amer. Math. Soc., Providence, R.I., 1979.

[17] Sturm, T.: On the Dirichlet-Poisson problem for Schrödinger operators. C.R. Math. Rep. Acad. Sci. Canada, Vol IX, No. 3 (1987), 149–154.

Heinz Leutwiler
Universität Erlangen – Nürnberg
Mathematisches Institut
Bismarckstr. 1 1/2
D-8520 Erlangen
F.R. Germany

B. Ja. Levin

Subharmonic Majorants and Some Applications

In this paper we shall briefly review the main facts referring to majorants (Perron envelopes) of special classes of functions, which are subharmonic in $\mathbb{C}$, and describe some applications of the majorants.

Some results in this area have been published, mainly without proofs [1–4]. The principal theorems were proved in preprints [5, 6][1]) . Applications to the theory of quasi-analytic classes are described in ref. [7] in detail.

The classes K_φ. By $\varphi(x)$ we denote an arbitrary non-negative function on $\mathbb{R}$, so that the value $\varphi(x) = +\infty$ is not excluded, and by K_φ any class of functions, subharmonic in $\mathbb{C}$, which satisfies the following conditions:

(a) all the functions $g(z)$ of the class K_φ satisfy for $x \in \mathbb{R}$ the inequality

$$g(x) \leq \varphi(x) ; \tag{1}$$

(b) if $g(z) \in K_\varphi$ and $g_c(z)$ is the function obtained from $g(z)$ by the balayage in a disk C, and if $g_c(x) \leq \varphi(x)$, then $g_c(z) \in K_\varphi$;

(c) if $g_1(z)$, $g_2(z) \in K_\varphi$, then also $g(z) = \max\{g_1(z), g_2(z)\} \in K_\varphi$;

(d) $0 \in K_\varphi$;

(e) there exists in K_φ a non-constant function[2]) ▲.

Below we shall only consider the symmetric classes K_φ, i.e. such ones that $g(z) \in K_\varphi$ implies $g(\bar{z}) \in K_\varphi$. The function $g(z)$, subharmonic in $\mathbb{C}$, will

---

[1]) P. Koosis [8, 9] developed the theory of superharmonic majorants of a different type as a method to obtain theorems similar to the widely known one due to A. Beurling and P. Malliavin on the multiplier [10].

[2]) The symbol ▲ signifies the end of a formulation.

be called a function of finite power if

$$\sigma = \varlimsup_{|z|\to\infty} \frac{g(z)}{|z|} < \infty, \tag{2}$$

and σ is the power of the function.

The class of all functions subharmonic in $\mathbb{C}$, of powers not larger than $\sigma \geq 0$, satisfying (1), will be denoted by K_φ^σ.

The symbol $K_{\varphi,\psi}$ will denote the class of all functions from K_φ° satisfying the condition

$$\varlimsup_{y\to+\infty} \frac{g(\pm iy)}{\psi(y)} < \infty \tag{3}$$

where $\psi(y) \uparrow \infty$. We also define $K_{\varphi,l} = K_{\varphi,\log|y|}$. ▲

Majorant of the class K_φ. Put $\hat{g}(z, K_\varphi) = \sup\limits_{g \in K_\varphi} \{g(z)\}$ and call the majorant of the class K_φ the function

$$V(z, K_\varphi) = \lim_{\delta\to 0} \left\{ \sup_{|\zeta - z| < \delta} \hat{g}(\zeta, K_\varphi) \right\}. \tag{4}$$

Sometimes, we shall write just $v(z)$.

Theorem 1. *The majorant $v(z)$ of the class K_φ is either equal to $+\infty$ in $\mathbb{C}$ or everywhere finite. In the latter case, $v(z)$ is harmonic in $\mathbb{C}_+(\mathbb{C}_-)$ and on those intervals of $\mathbb{R}$, where $v(x) < \varphi(x)$, may be represented as*

$$v(z) = \frac{|y|}{\pi} \int_{\mathbb{R}} \frac{v(t)}{|t-z|^2} dt + \sigma|y| \quad ▲. \tag{5}$$

This representation means that if the majorant of the class K_φ is finite, then this class is included in some class K_φ^σ. Besides, it follows from (5) that $v(z) = \sigma|y| + o(|z|)$ as $|z| \to \infty$ arbitrarily[3]) in $\mathbb{C}$.

By the complex majorant of the class K_φ we shall mean the function

$$w(z, K_\varphi) = u(z, K_\varphi) + iv(z, K_\varphi) \tag{6}$$

holomorphic in $\mathbb{C}_+(\mathbb{C}_-)$, or shorter $w(z) = u(z) + iv(z)$.

---

[3]) This asymptotic behaviour results from theorems of W.K. Hayman [11] and V.S. Azarin [12].

Theorem 2. *Let $\varphi(x) \geq 0$ and satisfy, for $x \in \mathbb{R}$, the following conditions:*

1) *the set of the points E, at which $\varphi(x)$ is finite, is closed;*

2) *the function $\varphi(x)$ is semi-continuous from below;*

3) *when $\forall x_0 \in E$, all the Lebesgue sets $E_{\varepsilon,x_0} = \{x : \varphi(x) \leq \varphi(x_0) + \varepsilon\}$, $\varepsilon > 0$ are closed and x_0 is a regular point of the boundary of the domain $\mathbb{C} \setminus E_{\varepsilon,x_0}$.*

Then, if the majorant of a given class K_φ is finite, the corresponding complex majorant is analytically continuable onto $\mathbb{R}$, $v(x) \leq \varphi(x)$ for $x \in \mathbb{R}$, the set $E_\varphi = \{x : \varphi(x) = v(x)\}$ is closed and the function $v(z)$ is harmonic on $\mathbb{C} \setminus E$.

The S-curve is defined as the continuous curve $u = u(x)$, $v = v(x)$, $x \in \mathbb{R}$ consisting of (a) the plot of the function $v = \theta(u)$, $-\infty \leq a < u < b \leq +\infty$, where $\theta(u)$ can only have first-kind discontinuity points and at these points $\theta(u_j) = \min\{\theta(u_j + 0), \theta(u_j - 0)\}$ and (b) the segments $h_j = \{w : \operatorname{Re} w = u_j, \; a_j \leq \operatorname{Im} w \leq b_j\}$, $j = 1, 2, \ldots$ where $a_j = \theta(u_j)$ and $b_j \geq \max\{\theta(u_j + 0), \theta(u_j - 0)\}$.

Ω_S will mean the domain bounded by the S-curve and lying above it.

The following theorem is central in the present topics.

Theorem 3. *The function $w = w(z)$ holomorphic in $\mathbb{C}_+$ is the complex majorant of a class K_φ, satisfying for φ the conditions of Theorem 2, if and only if it performs a conformal mapping of $\mathbb{C}_+$ onto a domain Ω_S.*

This theorem provides a method to solve some extremal problems in classes of subharmonic functions by reducing them to finding special conformal mappings (see refs. [1], [4], [5] and [13]). An E-regular mapping[4]) means here the conformal mapping of $\mathbb{C}_+$ onto the domain $\Omega(E)$, which results from dropping from the domain $-\infty \leq a < \operatorname{Re} w < b \leq +\infty$, $\operatorname{Im} w > 0$ a finite or a countable set of vertical segments starting at $\mathbb{R}$, having finite lengths and having no limit segments for $a < \operatorname{Re} w < b$. It is required that the given set $E = \bar{E} \subset \mathbb{R}$ should be transformed into $\mathbb{R}$ and the adjoint intervals, of which the set $\mathbb{R} \setminus E$ consists, into the vertical lines. The domain $\Omega(E)$ is evidently of the Ω_S-type, and by Theorem 3, the corresponding E-regular mapping is the complex majorant of a class K_φ, where $\varphi(x) = 0$ if $x \in E$, and $\varphi(x) = +\infty$ if $x \in \mathbb{R} \setminus E$.

---

[4]) *E-regular maps were studied in refs. [1], [2] and [5].*

Theorem 4. *If the set $E = \bar{E} \subset \mathbb{R}$ considered as the boundary of the domain $\mathbb{C} \setminus E$, is regular, then there exists the corresponding E-regular mapping. It is determined by the set E up to a similarity transformation and a real shift, i.e. $w(z) = \lambda w_0(z) + \mu$, $\lambda > 0$, $\mu \in \mathbb{R}$. Here $w(z)$ is an arbitrary E-regular mapping and $w_0(z)$ a fixed one[5]).*

Theorem 4 is the basis of the following classification of all regular sets $E \subset \mathbb{R}$ [1]. A set E belongs to the A-type if $a = -\infty$, $b = +\infty$, to the B-type if the quantity a or b is finite, and to the C-type if they both are finite. It is clear that if E belongs to the C-type, this means that it is "rare" in the vicinity of an infinitely distant point; and if it belongs to the B-type, it is highly asymmetrical. The type of the set is also related to growth. Here is such a theorem.

Theorem 5. *The set E belongs to the C-type if and only if $v(z) = O(\log|z|)$ as $|z| \to \infty$.*▲

In refs. [1] and [2], a different classification of closed sets was introduced. To describe it, let us note that by Theorems 1 and 3, the imaginary part of an E-regular map has a finite power. By Theorem 4, the imaginary parts of all E-regular maps, for a given set E, have either a positive power or zero power. Thus, all regular sets fall under two classes. A set E belongs to the class (α) if $v_E(z)$ has a positive power and to the class (β) if it has zero power. Clearly, the class (β) represents sets which are sufficiently rare in the vicinity of an infinitely distant point, while the class (α) are "massive" sets. The class (β) includes sets of all the three types A, B and C; and the class (α) only A-type sets.

Theorem 6. [1], [2], [6]. *If the set E belongs to the class (α), then the cone P_∞ of harmonic functions, positive in $\mathbb{C} \setminus E$ and equal to zero on E, is two-dimensional; and if it belongs to (β), then the cone P_∞ is one-dimensional[6]).*

This result was generalized by M. Benedicks [15] to positive harmonic functions in the space $\mathbb{R}^{n+1}$. Below we present one more condition for the set E to belong to the class (α).

---

[5]) This theorem was proved in ref. [6]. This proof is based on the well-known A. Pfluger theorem [14], which generalizes the Phragmén and Lindelöf theorems for an angle.

[6]) In terms of the Martin boundary, this theorem is formulated as follows: in the case (β) there is a single Martin point corresponding to an infinitely distant point, while in the case (α) there are two.

Theorem 7. *If E is a set relatively dense with respect to a measure, i.e. $\exists l > \delta > 0$ such that, for any $x \in \mathbb{R}$, mes $\{[x, x + l] \cap E\} > \delta$ then E belongs to the class (α). Besides, if σ is the power of the majorant $v_E(z)$, then for $x \in \mathbb{R}$*

$$v_E(x) \le \sigma C(l, \delta), \tag{7}$$

where $C(l, \delta)$ is a constant depending only on the above quantities.▲

Both statements of this theorem are essentially present in A.C. Schaeffer's paper [16] and also in refs. [1], [2] and [6]. In ref. [17] this theorem was generalized to polysubharmonic functions of many variables, and in the paper by M. Benedicks [15] to subharmonic functions in $\mathbb{R}^{n+1}$. A fairly general result of this character is the subject of ref. [18]. Recently A.E. Fryntov [19] found an exact value of the constant in inequality (7): $C(l, \delta) = \log \cot(\pi\delta/4l)$ and showed that it is only attained on the periodic set $E = \{x : nl \le x \le nl + \delta, \, n \in \mathbb{Z}\}$ (up to a shift of the set E).

E-regular maps have found an unexpected application in the spectral theory of the Sturm-Liouville differential operator with a periodic potential. The first result here was due to V.A. Marchenko and I.V. Ostrovski [20], [21]. They used E-regular maps to totally characterize the spectrum of the operator (in $L_2(-\infty, \infty)$)

$$\mathcal{L}y \overset{\text{def}}{=} -\ddot{y} + p(t)y = \lambda^2 y \tag{8}$$

where $p(t + 1) \equiv p(t)$ and $p(t) \in W_2^n(-\infty, \infty)$.

A certain transformation[7]) [22] can bring equation (8) to the form

$$y'' + \lambda^2 r(x)y = 0 \tag{9}$$

where $r(x) > 0$ and $r(x + \omega) = r(x)$.

M.G. Krein studied a somewhat more general equation, viz. the "string" equation:

$$\dot{y} + \lambda^2 \int_0^t y(\tau)d\sigma(\tau) = \text{const} \tag{10}$$

where $\sigma(t)$ is nondecreasing on $\mathbb{R}$ and $\sigma(t + 1) - \sigma(t) = \text{const}$ [23]; this equation however can also be reduced to a system of two differential equations

---

[7]) This transformation is: let $\lambda_0^2 < \lambda_1^2 < \cdots$ be the eigenvalues of the operator (8) under periodic boundary conditions. Without loss of generality, it may be assumed that $\lambda_0 = 0$. The corresponding first eigenfunction $\chi(t)$ is nowhere zero and $-\ddot{\chi} + p(t)\chi = 0$. Putting $z = y\chi^{-1}$, $x = \int_0^t \chi^{-2}(t)dt$ and $r(x) = \chi^4(t)$ we arrive at equation (9). In this case, $\omega = \int_0^1 \chi^{-2}(\tau)d\tau$.

with periodic coefficients. This reduction is carried out by the substitution $x = \sigma(t) + t$. The inverse function $t = \gamma(x)$ is defined as a constant on the intervals $\{\sigma(a_j - 0) + a_j, \sigma(a_j + 0) + a_j\}$. The functions $\gamma(x)$ and $\sigma(\gamma(x))$ obviously satisfy the Lipschitz condition and $0 \leq \gamma'(x) \leq 1$. By differentiating (10) with respect to x, we obtain $y' = \lambda\gamma'(x)z$, $z' = -\lambda(1 - \gamma'(x))y$.

By the substitution $y' = \lambda z$, equation (9) is obviously reduced to a system of two equations with coefficients from $L_\infty(0, \infty)$. More general is the system of two equations

$$Y' = \lambda J H(x) Y \tag{11}$$

where $Y = (y, z)$ is a two-dimensional vector function, $J = \begin{pmatrix} 0 & 1 \\ -1 & 0 \end{pmatrix}$, $H(x)$ is a real symmetrical matrix, $H(x + \omega) = H(x)$, $(H(x)\xi, \xi) > 0$, ξ being a two-dimensional vector, and the elements of the matrix $H(x)$ are absolutely integrable on $(0, \infty)$. The system (11) is called "canonical". Note that if the solution $Y = (y(x, \lambda), z(x, \lambda))$ of the system (11) is not identically equal to zero and $H(x) \neq 0$ on a set of a positive measure, then $\int_0^\omega (H(x)Y, Y)dx > 0$.

The spectrum, or the stability set means the set of the parameter values λ, for which there exists a solution of the system bounded on the whole $I\!\!R$.

Theorem 8. *The set E is a spectrum of the system* (11), *if and only if the corresponding E-regular mapping maps C_+ into a "regular comb", i.e. into the domain $\Omega_w \setminus \bigcup_n I_n$ where $\Omega_w = \{w : -\infty \leq a < \mathrm{Re}\, w < b \leq +\infty, \mathrm{Im}\, w > 0\}$ and $I_n = \{w : \mathrm{Re}\, w = nd, n \in \mathbb{Z}, 0 < \mathrm{Im} < h_n\}$.*

Theorem 9. *The set E is a spectrum of the string* (10), *if and only if the corresponding E-regular map transforms C_+ into a "regular comb" satisfying the additional condition $h_n = h_{-n}$. We mention refs.* [24], [25] *and* [26] *relevant to the problem.*

There is a close relation between the theorems on subharmonic majorants and the theory of quasi-analytic classes [2–4], [27]. This relation is thoroughly studied in ref. [7].

Suppose that $\varphi(x) \geq 0$ is a measurable function on the set E, $\int_E \exp(-\varphi(x))dx < \infty$ and denote by $F(E, \varphi)$ the class of functions

$$f(x) = \int\limits_E e^{i\lambda x} e^{-\varphi(\lambda)} \alpha(\lambda) d\lambda \tag{12}$$

where $\alpha(\lambda)$ is a bounded measurable function.

The curve $y = \varphi_1(x)$ is referred to as T_σ-convex, if it is the upper envelope of a family of ellipses $(x - \xi)^2 a^{-2} + y^2(a\sigma)^{-2} = 1$ for $a = a(\xi)$, $\xi \in I\!\!R$ and $\sigma > 0$ is a fixed number.

Theorem 10. *Let* (a) *the set* $E \subset \mathbb{R}$ *consist of the segments* $[a_k, b_k]$, $\cdots < b_{k-1} < a_k < b_k < a_{k+1} < \cdots$, $k \in \mathbb{Z}$ *and* $\inf\limits_{k}(b_k - a_k) > 0$; (b) *the curve* $y = \varphi_1(x)$ *be continuous and* T_σ*-convex, and* (c) *the function* $\varphi(x) = \varphi_1(x)$ *for* $x \in E$ *and* $\varphi(x) = +\infty$ *for* $x \in \mathbb{R} \setminus E$. *Then, for the class* $F(E, \varphi)$ *to be* I*-quasi-analytic, it is necessary and sufficient that the majorant* $v(z, K_\varphi^\sigma)$ *should be infinite.*

To explain this theorem, let us assume the set E and the function $\varphi(x)$ to satisfy the conditions (a), (b) and (c) of Theorem 10, define the function $\varphi_N(x) = \min(\varphi(x), N)$, $N > 0$ everywhere on $\mathbb{R}$ and span the maximal harmonic film, using those which "prop up" the line $y = \varphi_N(x)$ and satisfy the asymptotic condition $v(z, K_{\varphi_N}^\sigma) \sim \sigma|y|$ as $|y| \to \infty$. As N increases, the film rises. If the set E is "massive" and the function $\varphi(x)$ does not increase too fast, then the film tends to a limiting position $v(z, K_\varphi^\sigma)$.
If E is a "rare" set or the function $\varphi(x)$ increases fast as $x \to \infty$, then $v(z, K_{\varphi_N}^\sigma) \uparrow \infty$ i.e. $v(z, K_\varphi^\sigma) \equiv +\infty$ and the class $F(E, \varphi)$ is I-quasi-analytic. Note also that if $\varphi_1(x) = 0$, Theorem 10 can be formulated as

Theorem 11. *If the set* $E \subset \mathbb{R}$ *satisfies the condition* (a) *of Theorem* 10, *then the class* $F(E, 0)$ *is* I*-quasi-analytic if and only if the cone of positive functions, harmonic in* $\mathbb{C} \setminus E$, *that are zero on* E, *is one-dimensional.*

In conclusion, we present a theorem proving the relation between the Δ-quasi-analyticity and subharmonic majorants.

Assume that the set E satisfies the condition (a) of Theorem 10 and $\varphi(x)$ is an even function, so that $x\varphi'(x) \uparrow \infty$ as $x \to \infty$.

Theorem 12. *Under the above restrictions imposed on* E *and on* $\varphi(x)$, *the class* $F(E, \varphi)$ *is* Δ*-quasi-analytic if and only if* $v(z, K_{\varphi, l}) \equiv +\infty$.▲

Note that if the set E belongs to the class C, then the conditions of Theorem 12 are fulfilled and therefore the class $F(E, \varphi)$ is Δ-quasi-analytic, whatever the function $\varphi(x)$.

The general theory of majorants will appear in ref. [28–30].

References

[1] Akhiezer N.I. and Levin B.Ja. Generalizations of the Bernstein inequalities for derivatives of entire functions. In: "Studies in the Modern Problems of the Theory of Functions of a Complex Variable". Moscow, Fizmatgiz, 1960, pp. 125–183 (in Russian).

[2] Levin B.Ja. Subharmonic majorants and their applications. In: Abstracts of the All-Union Conference on the Theory of Functions, 1971, Kharkov, FTINT AN Ukr.SSR, pp. 117–120.

[3] Levin B.Ja. On some special conformal mappings. In: "Problems of Mathematics", coll. of papers No. 510, Tashkent, 1976, pp. 140–147 (in Russian).

[4] Levin B.Ja. Extremal problem in classes of subharmonic functions and their applications. Constructive theory of functions – 84. Sofia, 1984, pp. 534–543.

[5] Levin B.Ja. Majorants in classes of subharmonic functions and their applications. I. Preprint No. 18–84, FTINT AN Ukr.SSR, Kharkov 1984, pp. 1–52 (in Russian).

[6] Levin B.Ja. Majorants in classes of subharmonic functions and their applications. II. Preprint No. 19–84, FTINT AN Ukr.SSR, Kharkov 1984, pp. 1–34 (in Russian).

[7] Levin B.Ja. Completeness of a system of functions, quasi-analyticity and subharmonic majorants. Proc. of Steklov Inst. (Leningrad Department). Studies in the Linear Operator Theory and Function Theory, to appear.

[8] Koosis P. Fonctions entières de type exponentiel comme multiplicateurs. Un exemple et une condition nécessaire et suffisante. Ann. Scient. Ec. norm. sup., 4 série, t. 16, 1983, pp. 375–407.

[9] Koosis P. La plus petite majorante surharmonique ..., Annales de l'Institut Fourier de Grenoble, t. XXXII, Fasc. 1, 1983, pp. 67–107.

[10] Beurling A., Malliavin P. On Fourier transforms of measures with compact support. Acta Math., v. 107, 1962, pp. 291–309.

[11] Hayman W.K. Questions of regularity connected with the Phragmén-Lindelöf principle. J. Math. Pures Appl., 1956, v. 35, pp. 115–126.

[12] Azarin V.S. Generalization of a Hayman theorem to subharmonic functions in the n-dimensional cone. Mat. Sb., 1965, v. 66 (108), No. 2, pp. 248–264 (in Russian).

[13] Eremenko A.E. On the entire functions bounded on the real axis. DAN SSSR, 1987 (in Russian).

[14] Pfluger A. Des théorèmes du type de Phragmén-Lindelöf. C.R. Acad. Sci. Paris, t. 229, 1949, pp. 542–543.

[15] Benedicks M. Positive harmonic functions vanishing on the boundary of certain domain in $I\!R^n$. Ark. Math., 1980, v. 18, No. 1, pp. 53–72.

[16] Schaeffer A.C. Entire functions and trigonometric polynomials. Duke Math. J., 1953, v. 20, pp. 77–88.

[17] Katznelson V.E. Equivalent norms in spaces of entire functions of the exponential type. Mat. Sb., 1973, v. 92 (134), No. 1 (9), pp. 34–54 (in Russian).

[18] Levin B.Ja. and Logvinenko V.N. On the classes of the functions, subharmonic in $I\!R^n$ and bounded on a certain set. In: Proc. of Steklov Inst. (Leningrad Department), Coll. dedicated to Centenary of V.I. Smirnov (in Russian), to appear.

[19] Fryntov A.E. One extremal problem of the potential theorem. DAN SSSR (in Russian), to appear.

[20] Marchenko V.A. and Ostrovski I.V. Characterization of the spectrum of the Hill operator. Mat. Sb., 1975, v. 97 (139), No. 4 (8), pp. 540–606 (in Russian).

[21] Marchenko V.A. and Ostrovski I.V. Approximation of periodic potentials by finite-band ones. Vestnik Kharkovsk. Univ., No. 205. Priklad. Mat. i Mekhan., vyp. 45, 1980 (in Russian).

[22] Kovalenko K.R. and Krein M.G. On certain studies of A.M. Lyapunov in differential equations with periodic coefficients. DAN SSSR, v. XXV, No. 4, 1950, pp. 495–498 (in Russian).

[23] Krein M.G. On inverse problems of the theory of filters and stability λ-bands. DAN SSSR. v. 93, No. 5, 1953, pp. 767–770 (in Russian).

[24] Krein M.G. The main statements of the stability λ-bands of the canonical system of linear differential equations with periodic coefficients. A.A. Andronov Memorial Coll., Izd. AN SSSR, 1955, pp. 412–498 (in Russian).

[25] Misyura T.V. Characterization of the spectra of the periodic and antiperiodic boundary problems generated by the Dirac operation. I. The theory of functions, functional analysis and applications, 1978, vyp. 30, Kharkov, pp. 94–101; II. The theory of functions, functional analysis and applications, 1979, vyp. 31, Kharkov, pp. 102–109 (in Russian).

[26] Mikhailova I.V. The theory of the entire J-dilating matrix functions and its application in inverse problems. Synopsis of a Thesis, Kharkov, 1985 (in Russian).

[27] Kargaev P.P. Existence of the Phragmén-Lindelöf function and some conditions of quasi-analyticity. Proc. of Steklov Inst. (Leningrad Department), 1983, v. 126 (in Russian).

[28] Levin B.Ja. Majorants in classes of subharmonic functions. The theory of functions, functional analysis and applications, to appear in N51, Kharkov.

[29] Levin B.Ja. Connection between majorants and conformal mapping. Ibid., to appear in N52, Kharkov.

[30] Levin B.Ja. Classification of closed sets in $\mathbb{R}$ and representation of majorants. Ibid., to appear in N52, Kharkov.

Institute for Low Temperatures
Ukrainian Academy of Sciences
47 Lenin Prospect
Kharkov, 310164, USSR

Makoto Ohtsuka

On Weighted Extremal Length of Families of Curves

It is easy to see that the extremal length of order $p > 1$ of the family of curves terminating at one point (resp. tending to the point at infinity) is infinite if and only if $p \leq d$ (resp. $p \geq d$). In the present paper we generalize these results to the weighted extremal length.

§1 Preliminaries

In what follows we mean by a curve a non-point locally rectifiable curve in the Euclidean space R^d, $d \geq 2$. First we recall the definition of weighted extremal length of a familiy Γ of curves. We shall say that a Borel measurable function $\rho \geq 0$ in R^d is Γ-admissible if $\int_\gamma \rho ds \geq 1$ for every $\gamma \in \Gamma$. A measurable function $w \geq 0$ in R^d will be called a weight; functions equal to ∞ a.e. are not considered. We define the weighted module of order p of Γ by

$$M_p(\Gamma; w) = \inf \left\{ \int_{R^d} \rho^p w dx; \ \rho \text{ is } \Gamma\text{-admissible} \right\}$$

and the weighted extremal length by $\lambda_p(\Gamma; w) = 1/M_p(\Gamma; w)$. We shall write simply $\int$ for $\int_{R^d}$. This definition is a special case of the general one due to Fuglede [3].

We shall say that a curve family Γ is (p, w)-exc. if $\lambda_p(\Gamma; w) = \infty$ and that a property holds (p, w)-a.e. on a curve family if the exceptional curve family is (p, w)-exc.

Let $0 < \alpha < d$ and write $U_\alpha^\mu(x)$ (resp. $U_\alpha^f(x)$) for $\int |x-y|^{\alpha-d} d\mu(y)$ (resp. $\int |x-y|^{\alpha-d} f(y) dy$) whenever this has a meaning. Hereafter we let $p > 1$. We shall write $L^{p,w}$ for $\{f; \int |f|^p w dx < \infty\}$ and $L_+^{p,w}$ for $\{f \in L^{p,w}; f \geq 0\}$, and set

$$\| f \|_{p,w} = \left(\int |f|^p w dx \right)^{1/p}.$$

A set in R^d is called (α, p, w)-polar if there exists $f \in L_+^{p,w}$ such that $U_\alpha^f \not\equiv \infty$ and $U_\alpha^f = \infty$ on the set. It is known (cf. [3; Theorem 6]) that the ordinary extremal length, i.e. that with weight $w \equiv 1$, of order $p \leq d$ of the family $\Lambda(X)$ of curves terminating at points of a set $X \subset R^d$ is infinite if and only if X is $(1, p, 1)$-polar.

We shall say that a weight w satisfies the Muckenhoupt A_p condition ([5]) if

$$\sup_Q \frac{1}{|Q|} \int_Q w dx \left[\frac{1}{|Q|} \int_Q w^{1/(1-p)} dx \right]^{p-1} < \infty, \qquad (A_p)$$

where Q is a cube with sides parallel to the axes and $|Q|$ stands for the volume of Q. Then we write $w \in A_p$.

§2　Polar sets

We begin with

Theorem 1. *Let $0 < \alpha < d$. If a set X in R^d is (α, p, w)-polar, then*

$$\int_{|x|<R} (U_\alpha^\mu)^{p'} w^{1/(1-p)} dx = \infty \qquad (1)$$

for any measure $\mu \geq 0$, $\mu \not\equiv 0$, with $\operatorname{supp}\mu \subset X$ and for any $\{|x| < R\}$ meeting $\operatorname{supp}\mu$, where $1/p + 1/p' = 1$. Conversely, if X is an analytic set, $w^{1/(1-p)}$ is locally integrable and (1) holds for any measure μ and $\{|x| < R\}$ as above, then X is (α, p, w)-polar.

Proof. Suppose there exist $\mu \geq 0$, $\mu \not\equiv 0$, with $\operatorname{supp}\mu \subset X$ and $\{|x| < R\}$ meeting $\operatorname{supp}\mu$ such that the integral in (1) is finite. Take any $f \in L_+^{p,w}$ for which $U_\alpha^f \not\equiv \infty$. We have

$$\iint_{|x|<R} |x - y|^{\alpha-d} f(x) dx d\mu(y) = \int_{|x|<R} U_\alpha^\mu f dx$$

$$\leq \| f \|_{p,w} \left[\int_{|x|<R} (U_\alpha^\mu)^{p'} w^{1/(1-p)} dx \right]^{1/p'} < \infty.$$

Hence there exists $x^0 \in X \cap \{|x| < R\}$ such that $\int_{|x|<R} |x - x^0|^{\alpha-d} f(x) dx < \infty$. The fact $U_\alpha^f \not\equiv \infty$ implies $\int_{|x|>R} |x|^{\alpha-d} f(x) dx < \infty$. Since $\sup\{|x|/|x - x^0|;$ $|x| > R\}$ is finite, $\int_{|x|>R} |x - x^0|^{\alpha-d} f(x) dx < \infty$. Thus $U_\alpha^f(x^0) < \infty$ which shows that X is not (α, p, w)-polar.

Let us prove the converse under the assumptions stated above. Fix $R > 0$ so that $X \cap \{|x| < R\}$ is not empty. Write D_R for $\{|x| < R\}$ for simplicity. Set $\omega(x) = w(x)^{1/(1-p)}$ on D_R and $= 0$ elsewhere, and consider

$$C_{p,\omega}^{(\alpha)}(Y) = \inf\left\{\int f^p \omega dx; \quad f \geq 0, \ U_\alpha^{f\omega} \geq 1 \text{ on } Y\right\}$$

for any set $Y \subset R^d$ in case there exists such an f; otherwise set $C_{p,\omega}^{(\alpha)}(Y) = \infty$. Define also

$$c_{p,\omega}^{(\alpha)}(A) = \sup\left\{\mu(A); \ \mu \geq 0, \ \int (U_\alpha^\mu)^{p'} \omega dx \leq 1\right\}$$

for any analytic set A. Suppose $C_{p,\omega}^{(\alpha)}(X \cap D_R) > 0$. We apply Theorem 14 (ii) of [4] with $d\mu = \omega dx$ and obtain $c_{p,\omega}^{(\alpha)}(X \cap D_R) = (C_{p,\omega}^{(\alpha)}(X \cap D_R))^{1/p} > 0$. Therefore there exists μ such that $\mu(X \cap D_R) > 0$ and $\int (U_\alpha^\mu)^{p'} \omega dx \leq 1$. We find a compact set $K \subset X$ such that $\mu(K) > 0$ and denote the restriction of μ to K by μ_K. Then

$$1 \geq \int_{|x|<R} (U_\alpha^\mu)^{p'} \omega dx \geq \int_{|x|<R} (U_\alpha^{\mu_K})^{p'} \omega dx = \infty$$

by our assumption. This is impossible so that $C_{p,\omega}^{(\alpha)}(X \cap D_R) = 0$. Hence there exists $\{f_n\}$ such that $f_n \geq 0$, $U_\alpha^{f_n\omega} \geq 1$ on $X \cap D_R$ and $\| f_n \|_{p,\omega} < 2^{-n}$ for each n. We note that

$$\left[\int (f_n\omega)^p w dx\right]^{1/p} = \| f_n \|_{p,\omega} < 2^{-n}$$

and set $g = \sum_n f_n\omega$. Then $U_\alpha^g = \infty$ on $X \cap D_R$ and $\| g \|_{p,w} \leq 1$. Moreover, $U_\alpha^g \not\equiv \infty$ because

$$\int_{|x|<1} U_\alpha^g dx = \int_{|y|<R} g(y) dy \int_{|x|<1} \frac{1}{|x - y|^{d-\alpha}} dx$$

$$\leq \int_{|x|<1} \frac{dx}{|x|^{d-\alpha}} \| g \|_{p,w} \left[\int_{|y|<R} w^{1/(1-p)} dy\right]^{1/p'} < \infty.$$

Thus $X \cap D_R$ is (α, p, w)-polar. It is easy to conclude that X is (α, p, w)-polar. Our proof is completed. $\square$

Remark 1. The integral in (1) is always finite if $\alpha \geq d$, $U_\alpha^\mu \not\equiv \infty$ and $w^{1/(1-p)}$ is locally integrable.

Remark 2. The A_p condition is not assumed in the theorem.

Next we prove

Theorem 2. *Let* $w \in A_p$. *Then* X *is* $(1,p,w)$-*polar if and only if* $\lambda_p(\Lambda(X); w) = \infty$.

Before the proof we give some properties of weigthed extremal length. As in the case of ordinary extremal length we have $M_p(\Gamma; w) \leq M_p(\Gamma'; w)$ if $\Gamma \subset \Gamma'$ and $M_p(\cup_n \Gamma_n; w) \leq \sum_n M_p(\Gamma_n; w)$. We shall state some other properties as lemmas. For proofs see [3; Chap. I].

Lemma 1. *A family* Γ *of curves is* (p, w)-*exc. if and only if there exists* $\rho \in L_+^{p,w}$ *such that* $\int_\gamma \rho ds = \infty$ *for every* $\gamma \in \Gamma$.

Lemma 2. *If* E *is a Borel set of (Lebesgue) measure zero, then* $\int_{\gamma \cap E} ds = 0$ *for* (p, w)-*a.e.* γ.

Lemma 3. *If* $\| f_n \|_{p,w} \to 0$ *as* $n \to \infty$, *then there exists a subsequence* $\{n_k\}$ *such that* $\int_\gamma |f_{n_k}| ds \to 0$ *as* $k \to \infty$ *for* (p, w)-*a.e. curve* γ.

By a bounded curve we shall mean a curve which is bounded as a point set.

Lemma 4. *Let* w *be locally integrable and* Γ *be a family of bounded non-rectifiable curves. Then* $\lambda_p(\Gamma; w) = \infty$.

Proof. Let $\Gamma_n \subset \Gamma$ consist of curves contained in $|x| < n$. Then $\Gamma = \cup_n \Gamma_n$. Let $\rho = 1$ on $|x| < n$ and $= 0$ elsewhere. Evidently $\rho \in L_+^{p,w}$ and $\int_\gamma \rho ds = \infty$ for every $\gamma \in \Gamma_n$. Hence $M_p(\Gamma_n; w) = 0$ by Lemma 1 and $M_p(\Gamma; w) \leq \sum_n M_p(\Gamma_n; w) = 0$. This gives $\lambda_p(\Gamma; w) = \infty$. $\square$

Let $\Lambda_1(x)$ be the family of curves each of which contains at least one point of X. We shall prove

Lemma 5. *Assume that* w *is locally integrable. Then*

$$\lambda_p(\Lambda(X); w) = \lambda_p(\Lambda_1(X); w)$$

and they are 0 *or* ∞.

Proof. Let $\int_\gamma \rho ds \geq 1$ for all $\gamma \in \Lambda_1(X)$ and assume $\int \rho^p w dx < \infty$. If $\int_\gamma \rho ds < \infty$ for some $\gamma \in \Lambda_1(X)$, then there exists $\gamma' \in \Lambda_1(X)$ such that $\gamma' \subset \gamma$ and $\int_{\gamma'} \rho ds < 1$. This is impossible so that $\int_\gamma \rho ds = \infty$ for every $\gamma \in \Lambda_1(X)$. Lemma 1 implies $\lambda_p(\Lambda_1(X); w) = \infty$. Otherwise $\lambda_p(\Lambda_1(X); w) = 0$.

It is easy to see $\lambda_p(\Lambda(X); w) \leq \lambda_p(\Lambda_1(X); w)$. If $\lambda_p(\Lambda_1(X); w) = 0$, then $\lambda_p(\Lambda(X); w) = 0$. Assume $\lambda_p(\Lambda_1(X); w) = \infty$. Then there exists $\rho \in L_+^{p,w}$ such that $\int_\gamma \rho ds = \infty$ for every $\gamma \in \Lambda_1(X)$ by Lemma 1. Let $\Lambda' \subset \Lambda(X)$ be the family of curves each of which contains a bounded non-rectifiable curve of $\Lambda(X)$. Then $\lambda_p(\Lambda'; w) = \infty$ by Lemma 4. Next take any $\gamma \in \Lambda(X) - \Lambda'$. If γ is rectifiable and $\int_\gamma \rho ds < \infty$, then let $x_\gamma \in X$ be an end point of γ and consider the double $\hat{\gamma}$ of γ so that $x_\gamma \in \hat{\gamma}$. Then $\hat{\gamma} \in \Lambda_1(X)$ and $\int_{\hat{\gamma}} \rho ds < \infty$. This is impossible so that $\int_\gamma \rho ds = \infty$. If γ is not rectifiable, then any bounded subcurve γ' of γ, which terminates at a point of X, is rectifiable, $\int_\gamma \rho ds \geq \int_{\gamma'} \rho ds$ and $\int_{\gamma'} \rho ds = \infty$ as shown above. Hence $\int_\gamma \rho ds = \infty$ for every $\gamma \in \Lambda(X) - \Lambda'$. By Lemma 1 $\lambda_p(\Lambda(X) - \Lambda'; w) = \infty$ so that

$$M_p(\Lambda(X); w) \leq M_p(\Lambda'; w) + M_p(\Lambda(X) - \Lambda'; w) = 0.$$

Thus $\lambda_p(\Lambda(x); w) = \infty$. Therefore always $\lambda_p(\Lambda(X); w) = \lambda_p(\Lambda_1(X); w)$. $\square$

Lemma 6. *Let* $w \in A_p$, $f \in L^{p,w}$, $\varphi \in C_0^\infty$ *and* $\int \varphi dx = 1$. *Set* $\varphi_k(x) = k^d \varphi(kx)$. *Then* $f * \varphi_k \in C^\infty$ *and* $\| f * \varphi_k - f \|_{p,w} \to 0$ *as* $k \to \infty$.

Proof. It is easy to prove $f * \varphi_k \in C^\infty$. Let $\operatorname{supp} \varphi \subset \{|x| < R\}$. Then

$$|f * \varphi_k(x)| = k^d \left| \int_{|y-x|<R/k} f(y)\varphi(k(x - y))dy \right|$$

$$\leq \frac{R^d \max |\varphi|}{(R/k)^d} \int_{|y-x|<R/k} |f(y)|dy \leq \text{const.} \, Mf(x),$$

where Mf is the maximal function of f. Since $f * \varphi_k(x) \to f(x)$ as $k \to \infty$ for a.e. x and $\| Mf \|_{p,w} \leq \text{const.} \| f \|_{p,w}$ (cf. [5; Theorem 9]), the dominated convergence theorem yields that $\| f * \varphi_k - f \|_{p,w} \to 0$ as $k \to \infty$. $\square$

Remark. The original proof by the author was simplified as above by Aikawa; See [1; Lemma 4].

Proof of Theorem 2. Suppose X is $(1, p, w)$-polar. There exists $f \in L_+^{p,w}$ such that $U_1^f \not\equiv \infty$ and $U_1^f = \infty$ on X. Consider φ_k as in Lemma 6. According to it $f * \varphi_k \in C^\infty$ and $\| f * \varphi_k - f \|_{p,w} \to 0$ as $k \to \infty$. We have

$$\| \operatorname{grad} (U_1^{f*\varphi_k} - U_1^f) \|_{p,w} \leq \text{const.} \ \| f * \varphi_k - f \|_{p,w} \to 0$$

as $k \to \infty$ by [2]. By Lemma 3 there exist a subsequence $\{k_\ell\}$ and a (p, w)-exc. family Γ_1 of curves such that $\int_\gamma |\operatorname{grad} U_1^{f*\varphi_{k_\ell}} - \operatorname{grad} U_1^f| ds \to 0$ as $\ell \to \infty$ if $\gamma \notin \Gamma_1$. For simplicity we write f_ℓ for $f * \varphi_{k_\ell}$. It follows that $\int_\gamma |\operatorname{grad} (U_1^{f_\ell} - U_1^{f_m})| ds \to 0$ as $\ell, m \to \infty$. Let E be the set of points x at which $U_1^{f*\varphi_k}(x) = U_1^f * \varphi_k(x) \to U_1^f(x)$ as $k \to \infty$. One can see that $R^d - E$ is of measure zero. Let $B \supset R^d - E$ be a Borel set of measure zero. By Lemma 2 there exists a (p, w)-exc. family Γ_2 of curves such that $\int_{\gamma \cap B} ds = 0$ if $\gamma \notin \Gamma_2$. For $\gamma \notin \Gamma_1 \cup \Gamma_2$ take a point $x^0 \in \gamma - B$. For any $x \in \gamma$ we have

$$|U_1^{f_\ell}(x) - U_1^{f_m}(x)| \leq |U_1^{f_\ell}(x^0) - U_1^{f_m}(x^0)| + \int_\gamma |\operatorname{grad} (U_1^{f_\ell} - U_1^{f_m})| ds \to 0$$

as $\ell, m \to \infty$. Therefore $U_1^{f_\ell}$ converges at every point of γ. At every point ξ of X, $U_1^{f*\varphi_k}(\xi) \to U_1^f(\xi) = \infty$ as $k \to \infty$ because U_1^f is lower semicontinuous. Accordingly $U_1^f < \infty$ on γ. Namely, $\gamma \notin \Lambda_1(X)$ and hence $\Lambda_1(X) \subset \Gamma_1 \cup \Gamma_2$, which gives $\lambda_p(\Lambda_1(X); w) \geq \lambda_p(\Gamma_1 \cup \Gamma_2; w)$. Since both Γ_1 and Γ_2 are (p, w)-exc., $\lambda_p(\Gamma_1 \cup \Gamma_2; w) = \infty$ so that $\lambda_p(\Lambda_1(X); w) = \infty$. By Lemma 5 we obtain $\lambda_p(\Lambda(X); w) = \infty$. The only-if part of the theorem is thus proved.

Conversely, assume $\lambda_p(\Lambda(X); w) = \infty$. By Lemma 1 there exists $h \in L_+^{p,w}$ such that $\int_\gamma h ds = \infty$ for every $\gamma \in \Gamma(X)$. Set $h_1(x) = (1 + |x|^2)^{-1/2} h(x)$. Evidently $h_1 \in L_+^{p,w}$. By Lemma 3 of [1]

$$\int h_1(x) (1 + |x|^2)^{(1-d)/2} dx = \int h(x) (1 + |x|^2)^{-d/2} dx < \infty.$$

Hence $U_1^{h_1} \not\equiv \infty$. Suppose there exists $x^0 \in X$ with $U_1^{h_1}(x^0) < \infty$. Using the polar coordinates at x^0, we have

$$\infty > U_1^{h_1}(x^0) = \iint h_1 dr d\Theta.$$

Therefore $\int h_1 ds < \infty$ along a.e. ray issuing from x^0. Let γ be a ray such that $\int_\gamma h_1 ds < \infty$ and $\gamma' = x^0 a$ be the segment on γ of length 1. Evidently $\gamma' \in \Lambda(X)$. We have

$$\int_{\gamma'} h ds = \int_{\gamma'} (1 + |x|^2)^{1/2} h_1 ds < (2 + |x^0|)^{1/2} \int_{\gamma'} h_1 ds < \infty.$$

This is contradictory to the fact that $\int h\,ds = \infty$ along every curve of $\Lambda(X)$. Thus $U_1^{h_1} = \infty$ on X. It is shown that X is $(1, p, w)$-polar. The proof of our theorem is complete. $\square$

Taking the origin as the set X in Theorems 1 and 2 we derive

Corollary. *Let $w \in A_p$. Then $\lambda_p(\Lambda(\{0\}); w) = \infty$ if and only if*

$$\int_{|x|<1} |x|^{(1-d)p'} w(x)^{1/(1-p)}\,dx = \infty. \tag{2}$$

Remark. We can show that $\lambda_p(\Lambda(\{0\}); w) = \infty$ implies (2), without assuming the A_p condition. The proof is similar to that of the first half of Theorem 3.

§3 Point at infinity

The last theorem is

Theorem 3. *If $\lambda_p(\Lambda_\infty; w) = \infty$, then*

$$\int_{|x|>1} |x|^{(1-d)p'} w(x)^{1/(1-p)}\,dx = \infty, \tag{3}$$

where Λ_∞ is the family of curves tending to the point at infinity. Conversely, if $w \in A_p$ and (3) holds, then $\lambda_p(\Lambda_\infty; w) = \infty$.

Proof. Assume that the integral in (3) is finite. Let Γ be the family of rays $\{r\Theta;\ 1 < r,\ |\Theta| = 1\}$. Evidently $\Gamma \subset \Lambda_\infty$. Let ρ be Γ-admissible. Then

$$1 \leq \int_\gamma \rho\,ds = \int_1^\infty \rho(r, \Theta)\,dr$$

for any ray γ issuing from a point Θ with $|\Theta| = 1$. We have

$$\int_{|\Theta|=1} d\Theta \leq \int_{|\Theta|=1} \int_1^\infty \rho r^{1-d} r^{d-1}\,dr\,d\Theta = \int_{|x|>1} \rho |x|^{1-d}\,dx$$

$$\leq \| \rho \|_{p,w} \left[\int_{|x|>1} |x|^{(1-d)p'} w^{1/(1-p)}\,dx \right]^{1/p'}$$

so that

$$\| \rho \|_{p,w}^p \geq \left[\int_{|\Theta|=1} d\Theta \right]^p \left[\int_{|x|>1} |x|^{(1-d)p'} w^{1/(1-p)} dx \right]^{-p/p'} > 0$$

by our assumption. Thus $0 < M_p(\Gamma; w) \leq M_p(\Lambda_\infty; w)$ which implies $\lambda_p(\Lambda_\infty; w) < \infty$.

Conversely, assume $w \in A_p$ and (3). Assume moreover that $\int_{|x|>1} |x|^{1-d} f dx < \infty$ for any $f \in L_+^{p,w}$. Then there exists a constant c such that

$$\int_{|x|>1} |x|^{1-d} f dx \leq c \| f \|_{p,w}$$

for any $f \in L_+^{p,w}$. The function

$$f_j(x) = \chi_{\{1<|x|<j\}} |x|^{(1-d)(p'-1)} w(x)^{1/(1-p)}$$

belongs to $L_+^{p,w}$ so that

$$\int_{|x|>1} |x|^{1-d} f_j dx = \int_{1<|x|<j} |x|^{(1-d)p'} w^{1/(1-p)} dx$$

$$\leq c \left[\int_{1<|x|<j} \left(|x|^{(1-d)(p'-1)} w^{1/(1-p)} \right)^p w dx \right]^{1/p}$$

$$= c \left[\int_{1<|x|<j} |x|^{(1-d)p'} w^{1/(1-p)} dx \right]^{1/p}$$

which gives

$$\int_{1<|x|<j} |x|^{(1-d)p'} w^{1/(1-p)} dx \leq c^{p'}.$$

Letting $j \to \infty$ we obtain

$$\int_{|x|>1} |x|^{(1-d)p'} w^{1/(1-p)} dx \leq c^{p'}$$

contrary to (3). Consequently $\int_{|x|>1} |x|^{1-d} f dx = \infty$ for some $f \in L_+^{p,w}$.

Now take R_n so that $1-1/n < |y|^{d-1} |x-y|^{1-d} < 1+1/n$ for $|x| < n$ and $|y| > R_n$. There exists $\psi \in C_0^\infty$ such that $0 \leq \psi \leq 1$, $\operatorname{supp} \psi \subset \{y; |y| > R_n\}$ and $g = \psi f$ with the above f satisfies $U_1^g(0) = 1$ and $\| g \|_{p,w} < 1/n$. We observe that $1-1/n < U_1^g(x) < 1+1/n$ or $|1-U_1^g(x)| < 1/n$ for x with $|x| < n$.

Let $\varphi \in C_0^\infty$, $\varphi \geq 0$, $\operatorname{supp}\varphi \subset \{|x| < 1\}$, $\int \varphi dx = 1$ and $\varphi_k(x) = k^d \varphi(kx)$. We note that $g * \varphi_k \in C_0^\infty$ and $U_1^g * \varphi_k = U_1^{g*\varphi_k} \in C^\infty$. Moreover, $g * \varphi_k \to g$ in $L^{p,w}$ by Lemma 6 so that $\| g * \varphi_k \|_{p,w} \to \| g \|_{p,w} < 1/n$ as $k \to \infty$. By the aid of Theorem III of [2] we have

$$\| \operatorname{grad} U_1^{g*\varphi_k} \|_{p,w} \leq \text{const.} \| g * \varphi_k \|_{p,w} < \text{const.}/n$$

for a large k, say for $k = k_n \geq 2$. We observe also that

$$|1 - U_1^{g*\varphi_k}(x)| = |1 - U_1^g * \varphi_k(x)| = |(1 - U_1^g) * \varphi_k(x)| < 1/n$$

for x on $|x| < n/2$. Write g_n for $g * \varphi_{k_n}$ for simplicity. Setting $\rho_n = |\operatorname{grad} U_1^{g_n}|$ we have $\| \rho_n \|_{p,w} < \text{const.}/n$. Let $\Gamma_j \subset \Lambda_\infty$ be the family of curves each of which meets $\{|x| < j/2\}$. Then

$$\int_\gamma \rho_n ds \geq U_1^{g_n}(x_\gamma) \geq 1 - 1/n$$

for every $\gamma \in \Gamma_j$ if $j \leq n$, where x_γ is a point on γ contained in $\{|x| < n/2\}$. Hence

$$M_p(\Gamma_j; w) \leq (1 - 1/n)^{-p} \int \rho_n^p w dx \leq \frac{\text{const.}}{(n-1)^p} \to 0 \quad \text{as} \quad n \to \infty.$$

Thus $M_p(\Gamma_j; w) = 0$ for each j so that $M_p(\Lambda_\infty; w) = 0$ because $\Lambda_\infty = \cup_j \Gamma_j$. Our theorem is now proved. $\square$

Remark 1. Aikawa gave various characterizations of the condition

$$\int_{|x|>1} |x|^{(m-d)p'} w(x)^{1/(1-p)} dx < \infty$$

in [1; Theorem 3].

Remark 2. Let Γ be the family of rays as above. Writing $w(x) = w(r, \Theta)$ and setting $h(\Theta) = \int_1^\infty (r^{d-1} w(r, \Theta))^{1/(1-p)} dr$, we derive

$$M_p(\Gamma; w) = \int_{|\Theta|=1} h^{1-p}(\Theta) d\Theta$$

from Theorem 1 of [6]. Hence $\lambda_p(\Gamma; w) = \infty$ implies $h(\Theta) = \infty$ for a.e. Θ. Since the integral in (3) is equal to $\int_{|\Theta|=1} h(\Theta) d\Theta$, condition (3) follows. Thus an alternative proof of the first half of Theorem 3 is obtained.

Remark 3. Let Γ and h be the same. There exists w for which $\int_{|\Theta|=1} h(\Theta)d\Theta$ $= \infty$ but $\int_{|\Theta|=1} h^{1-p}(\Theta)d\Theta > 0$; for instance, take $w(r,\Theta) = r^{p-d}$ for Θ of a semisphere on $|\Theta| = 1$ and $= r^{2p-d-1}$ for the other Θ's. Then (3) is true but $\lambda_p(\Lambda_\infty; w) \leq \lambda_p(\Gamma; w) < \infty$. This shows that the A_p condition is necessary in the latter half of Theorem 3. In such a way we obtain some examples of w which do not satisfy the A_p condition.

Remark 4. Let Γ and h be the same and assume $w \in A_p$. If $\int_{|\Theta|=1} hd\Theta = \infty$, then $M_p(\Lambda_\infty; w) = 0$ by Theorem 3. Since $M_p(\Gamma; w) \leq M_p(\Lambda_\infty; w)$,

$$0 = M_p(\Gamma; w) = \int_{|\Theta|=1} h^{1-p}d\Theta$$

which implies $h(\Theta) = \infty$ for a.e. Θ. As shwon in Remark 3 this is not always true if $w \notin A_p$. Thus the A_p condition seems to indicate a kind of rotation free character of $h(\Theta)$.

Remark 5. We can state similar remarks in relation to the Corollary to Theorem 2 and the Remark following it.

References

[1] H. Aikawa: On weighted Beppo Levi functions, in preparation.

[2] R.R. Coifman and C. Fefferman: Weighted norm inequalities for maximal functions and singular integrals, Studia Math. 51 (1974), 241–250.

[3] B. Fuglede: Extremal length and functional completion, Acta Math. 98 (1957), 171–219.

[4] N.G. Meyers: A theory of capacities for potentials of functions in Lebesgue classes, Math. Scand. 26 (1970), 255–292.

[5] B. Muckenhoupt: Weighted norm inequalities for the Hardy maximal function, Trans. Amer. Math. Soc. 165 (1972), 207–226.

[6] M. Ohtsuka: Extremal length of level surfaces and orthogonal trajectories, J. Sci. Hiroshima Univ. Ser. A–I 28 (1964), 259–270.

Department of Mathematics
Gakushuin University
Mejiro, Toshima-ku
Tokyo, 171 Japan

Edgar Reich

On Approximation by Rational Functions of Class L^1

§1 Fundamental results

Let $S = \{z_k\}$, $0 \leq |z_1| \leq |z_2| \leq \ldots$, be a countably infinite set in the complex plane $\mathbb{C}$ with no limit points in $\mathbb{C}$. We denote by $\mathcal{B}_S$ the collection of functions $f(z)$, analytic in $\mathbb{C} \setminus S$, possessing finite L^1 norm,

$$\| f \| = \iint_{\mathbb{C}} |f(z)| dx dy < \infty \qquad (z = x + iy).$$

The class $\mathcal{B}_S$ plays a basic role in connection with certain extremal problems for quasiconformal mappings ([3], [4]). For orientation we recall the following corollary of a general result of Bers [1]. (A constructive proof is given in [2].)

Theorem A. *Given $f \in \mathcal{B}_S$, there exists a sequence $\{f_n\}$ of rational functions of class $\mathcal{B}_S$, such that*

(i) $\lim f_n(z) = f(z)$ *uniformly in every compact subset of* $\mathbb{C} \setminus S$

(ii) $\lim \| f - f_n \| = 0.$

Since it is easy to see that every isolated singularity of a member of $\mathcal{B}_S$ is a simple pole, statement (ii) identifies the L^1-*closure* of the class of rational functions in $\mathcal{B}_s$, up to L^1-equivalence, as a subset of the set of meromorphic functions whose poles lie on S and are of order at most 1. The closure of the same class with respect to *local uniform convergence* is of course at least as large; in fact, it turns out to be considerably larger, and to contain e.g. the set of all entire functions, while, by contrast, the L^1-closure contains only the single entire function 0. We shall prove the following:

Theorem 1. *There exists a sequence $\{f_n\}$ of rational functions of class $\mathcal{B}_S$, such that*

$$\lim f_n(z) = f(z), \quad loc.\ unif.\ in\quad \mathbb{C} \setminus S,$$

if and only if $f(z)$ is meromorphic in $\mathbb{C}$ and all poles of f are simple and lie on S.

Proof. If such a sequence $\{f_n\}$ exists, then since $\| f_n \| < \infty$, $(n = 1, 2, \ldots)$, f_n has poles of at most order 1; of course, they can only be located at points z_k of S. Since $(z - z_k)f_n(z)$ is analytic at z_k, it follows by uniform convergence on a small circle, $|z - z_k| = \epsilon$, that f has at worst a simple pole at z_k. Thus, necessity is proved.

To prove sufficiency, suppose $f(z)$ has residue μ_k at $z = z_k$. We can represent $f(z)$ à la Mittag-Leffler as

$$f(z) = \sum_{k=1}^{\infty} \left[\frac{\mu_k}{z - z_k} - P_k(z) \right] + \sum_{m=0}^{\infty} A_m z^m, \quad \text{loc. unif. in} \quad \mathbb{C} \setminus S, \quad (1.1)$$

where $P_1(z) = 0$, and $P_k(z)$, $(k \geq 1)$, is a polynomial obtained by truncating the power series of $\mu_k(z - z_k)^{-1}$, $(|z| < |z_k|)$; the power series $\sum A_m z^m$ is an entire function.

The most general rational function belonging to $\mathcal{B}_S$ has the form of a *finite* sum

$$\sum_k \frac{\nu_k}{z - z_k},$$

whose coefficients satisfy

$$\sum_k \nu_k = \sum_k z_k \nu_k = 0,$$

in order for the L^1-norm to be finite. In particular, if u, v, w are distinct points of S, then the rational function

$$R(u, v, w; z) = \frac{1}{z - u} - \frac{w - u}{w - v} \frac{1}{z - v} + \frac{v - u}{w - v} \frac{1}{z - w} \quad (1.2)$$

belongs to $\mathcal{B}_S$.

Since every finite linear combination of rational functions in $\mathcal{B}_S$ with complex coefficients is also a rational function in $\mathcal{B}_S$, it suffices, in view of (1.1), to show that $(z - u)^{-1}$, $(u \in S)$, the constant 1, and z^m, $(m = 1, 2, \ldots,)$, are each representable as local uniform limits of rational functions in $\mathcal{B}_S$. We now proceed to do this.

(a) If we choose $w_n \in S$, $n = 1, 2, \ldots$, such that

$$|z_n/w_n| \to 0,$$

then

$$R(u, z_n, w_n; z) \to \frac{1}{z - u}, \quad \text{loc. unif. in} \quad \mathbb{C} \setminus S.$$

(b) Choose $u_n, v_n, w_n \in S$, $n = 1, 2, \ldots$, such that

$$u_n/v_n \to 0, \quad w_n/v_n \to 0.$$

Then

$$-u_n R(u_n, v_n, w_n; z) \to 1, \quad \text{loc. unif. in} \quad \mathbb{C} \setminus S.$$

(c) We consider

$$\Delta_n(z) = \alpha_n R(\alpha_n, \beta_n, \gamma_n; z) - u_n R(u_n, v_n, w_n; z),$$

where, given $u_n \in S$, we firstly choose $\alpha_n \in S$, $(n = 1, 2, \ldots)$, such that

$$|\alpha_n| > 2|u_n|, \qquad (n = 1, 2, \ldots,). \tag{1.3}$$

Since

$$\frac{u_n \alpha_n}{\alpha_n - u_n} \Delta_n(z) - \frac{u_n \alpha_n z}{(z - u_n)(z - \alpha_n)}$$

$$= \frac{u_n \alpha_n}{\alpha_n - u_n} \left\{ -\frac{\gamma_n - \alpha_n}{\gamma_n - \beta_n} \frac{\alpha_n}{z - \beta_n} + \frac{\beta_n - \alpha_n}{\gamma_n - \beta_n} \frac{\alpha_n}{z - \gamma_n} + \right.$$

$$\left. + \frac{w_n - u_n}{w_n - v_n} \frac{u_n}{z - v_n} - \frac{v_n - u_n}{w_n - v_n} \frac{v_n - u_n}{w_n - v_n} \frac{u_n}{z - w_n} \right\},$$

setting

$$g_n(z) = \frac{u_n \alpha_n}{\alpha_n - u_n} \Delta_n(z),$$

it follows, by (1.3), that

$$\left| g_n(z) - \frac{u_n \alpha_n z}{(z - u_n)(z - \alpha_n)} \right| \le \left| \frac{\gamma_n - \alpha_n}{\gamma_n - \beta_n} \right| \frac{|\alpha_n^2|}{|z - \beta_n|} + \frac{|\beta_n - \alpha_n|}{|\gamma_n - \beta_n|} \frac{|\alpha_n^2|}{|z - \gamma_n|} \right\}$$

$$+ \left| \frac{w_n - u_n}{w_n - v_n} \right| \frac{|u_n \alpha_n|}{|z - v_n|} + \frac{|v_n - u_n|}{|w_n - v_n|} \frac{|u_n \alpha_n|}{|z - w_n|}. \tag{1.4}$$

Next, we choose $\beta_n \in S$, $v_n \in S$, such that

$$\alpha_n^2/\beta_n \to 0, \quad (\alpha_n u_n)/v_n \to 0, \tag{1.5}$$

and finally we choose $\gamma_n \in S$, $w_n \in S$, such that

$$\beta_n/\gamma_n \to 0, \quad v_n/w_n \to 0. \tag{1.6}$$

As a consequence of (1.5), (1.6), we are automatically assured that

$$\alpha_n/\gamma_n \to 0, \quad u_n/w_n \to 0. \tag{1.7}$$

Therefore, the right-hand side of (1.4) goes to zero as $n \to \infty$, loc. unif. in $\mathbb{C} \setminus S$. Thus,

$$g_n(z) \to z, \quad \text{loc. unif. in} \quad \mathbb{C} \setminus S.$$

(d) We observe the following: If $f_1(z)$ and $f_2(z)$ are rational functions with first-order poles at distinct points, and if

$$z^2 f_1(z) \to 0, \quad z^2 f_2(z) \to 0, \quad \text{as} \quad z \to \infty,$$

then the product $f_1(z)f_2(z)$ has these same properties; that is, the point-wise *product* of rational functions of class $\mathcal{B}_S$ *without common poles* is also a rational function of class $\mathcal{B}_S$.

In order to represent z^2 we choose two disjoint countably infinite sub-sequences S_1, S_2, of S, and form $\{g_{1n}(z)\}$ using S_1, and $\{g_{2n}(z)\}$ using S_2, in the same manner that $\{g_n(z)\}$ was formed using S in part (c). Then $g_{1n}(z)g_{2n}(z)$ is a rational function in $\mathcal{B}_S$, and

$$\lim_{n \to \infty} [g_{1n}(z)g_{2n}(z)] = z^2, \quad \text{loc. unif. in} \quad \mathbb{C} \setminus S.$$

As is readily seen, representations of $z^3, z^4, \ldots$, as limits of rational functions in $\mathcal{B}_S$ can be obtained inductively.

This completes the proof. $\square$

Suppose $X = \{x_k\}$ is a countably infinite set of real numbers, $|x_1| \leq |x_2| \leq \ldots$, $\lim |x_k| = \infty$. By examining the proof of Theorem 1, it is easily verified that Theorem 1 has the following symmetric version:

Theorem 2. *There exists a sequence $\{\phi_n\}$ of rational functions of class $\mathcal{B}_X$ satisfying $\phi_n(\bar{z}) = \overline{\phi_n(z)}$, such that*

$$\lim \phi_n(z) = \phi(z), \quad \text{loc. unif. in} \quad \mathbb{C} \setminus X,$$

if and only if $\phi(z)$ is meromorphic in $\mathbb{C}$, possesses no poles other than simple poles lying in X, and satisfies $\phi(\bar{z}) = \overline{\phi(z)}$.

§2 Construction of extremal Teichmüller mappings — a counterexample

Following [4], let us consider the upper half-plane,

$$H = \{z : \operatorname{Im} z > 0\}$$

together with a countably infinite collection $X = \{x_k\}$, $|x_1| \leq |x_2| \leq \ldots,$ of isolated distinguished boundary points ("vertices") on the real axis. The configuration (H, X) is referred to as a *polygon* P with vertices X. Suppose $X_n = \{x_1, x_2, \ldots, x_n\}$, $(n \geq 4)$, is a finite subset of vertices of P. We shall say that a function $g(z)$ is *admissible* with respect to the polygon P, if g is a rational function whose poles are restricted to X_n for some n, if the residues at the poles are all real, and if the L^1 norm of g over H is finite. Evidently, g is admissible with respect to P if and only if g is a rational function of class $\mathcal{B}_X$, and $g(\bar{z}) = \overline{g(z)}$.

Suppose P', with vertices $\{x'_k\}$, is another polygon, quasiconformally equivalent to P in the sense that a quasiconformal self-mapping of H taking x_k onto x'_k, $(k = 1, 2, \ldots,)$ and ∞ onto ∞ exists. A known procedure for constructing an *extremal* quasiconformal mapping of P onto P' (one minimizing the maximal dilatation) involves the following. One associates to X_n a "Teichmüller" extremal mapping f_n with complex dilatation

$$\frac{f_{n\bar{z}}}{f_{nz}} = k_n \frac{\overline{\phi_n(z)}}{|\phi_n(z)|}, \quad z \in H. \tag{2.1}$$

The above functions $\phi_n(z)$ are chosen so as to be admissible with respect to P, and in such a manner that

$$f_n(H) = H, \quad f_n(x_k) = x'_k, \quad (k = 1, 2, \ldots, n), \quad f_n(\infty) = \infty. \tag{2.2}$$

In order to avoid triviality we can assume that

$$\lim k_n = k_0 > 0.$$

Since $\{f_n(z)\}$ is a normal family, the existence of a subsequence $\{f_{n_j}\}$ converging locally uniformly to a mapping f_0 of P onto P' is guaranteed, and it is straightforward to verify that f_0 is an extremal mapping of P onto P'. It is of interest to know whether, additionally, $\{\phi_n\}$ possesses a subsequence $\{\phi_{n_j}\}$ such that

$$\frac{\overline{\phi_{n_j}(z)}}{|\phi_{n_j}(z)|} \to \frac{\overline{\phi_0(z)}}{|\phi_0(z)|} \quad \text{a.e. in} \quad H, \tag{2.3}$$

where $\phi_0(z)$ is analytic in H. If (2.3) is the case, one could conclude that there exists an extremal mapping f_0 of P onto P', possessing a complex dilatation of the form

$$k_0 \frac{\overline{\phi_0(z)}}{|\phi_0(z)|}, \quad z \in H;$$

i.e., f_0 would be a Teichmüller mapping corresponding to a quadratic differential ϕ_0 in H of possibly infinite L^1 norm.

From [4], a question that implicitly suggests itself is whether from the fact alone that $\{\phi_n\}$ are admissible with respect to P (i.e., neglecting the additional information that (2.2) is satisfied), one could conclude that a subsequence satisfying (2.3) exists. We can now answer this in the *negative* in the following strong sense:

For any polygon P there exists a sequence of admissible rational functions $\{\phi_m(z)\}$ such that no subsequence of the sequence $\{\overline{\phi_m(z)}/|\phi_m(z)|\}$ converges on a set of positive measure.

Proof. Applying Theorem 2 with $\phi(z) = e^{mz}$ we can find a rational function $\phi_m(z)$ of class $\mathcal{B}_X$ with real residues, such that

$$|\phi_m(z) - e^{mz}| < \frac{1}{2^m} \quad \text{for} \quad |z| < m.$$

The sequence $\{\phi_m(z)\}$ has the property claimed. $\square$

Whether indeed an extremal Teichmüller mapping f_0 of P onto P' exists, and to what extent the existence depends on the choice of P and P' is still an open question. The foregoing shows that more information about $\{\phi_n\}$ than merely admissibility with respect to P is needed to establish the existence of such an f_0.

References

[1] L. Bers, "An approximation theorem", J. Analyse Math. *14* (1965), 1–4.

[2] Edgar Reich, "L^1-approximation of meromorphic functions", J. Approximation Theory *31* (1981), 1–5.

[3] Edgar Reich and Kurt Strebel, "Quasiconformal mappings of the punctured plane", Springer Lecture Notes in Math., *103* (1983), 182–212.

[4] Kurt Strebel, "On the existence of extremal Teichmüller mappings", Complex Variables *9* (1987), 287–295.

Forschungsinstitut für Mathematik   and   University of Minnesota,
ETH-Zürich   Minneapolis

H. Renggli

On Fixed Points of Conformal Automorphisms of Riemann Surfaces

0. Let R be a compact Riemann surface of genus g, and let f denote a conformal automorphism of R that is not the identity mapping. By a classical theorem the number of fixed points of f is at most $2g + 2$. This result has in recent years been generalized to non-compact Riemann surfaces (cf. [P/L] for $g = 0$ and [M], [S] for $g \geq 0$). Our paper contains new proofs of that fact.

Let R be a hyperbolic Riemann surface that is not simply connected. It is shown first that for any two distinct fixed points of f there are always more than one hyperbolic geodesic that connects them (Proposition 1). Using that fact one easily concludes that the number of fixed points is at most 2 if $g = 0$ (Proposition 2). Next a result of H.M. Farkas [F] for compact Riemann surfaces is generalized to non-compact ones (Proposition 3). It then follows that $2g + 2$ is the maximal number of fixed points if $g > 0$ (Corollary).

1. We shall use the **Notations.**

R: a hyperbolic Riemann surface that is not simply connected. (The universal covering surface U of R is mapped by ρ conformally onto the unit disc D and by the projection mapping π onto R.)

g: the genus of R (assumed to be finite).

f: a conformal automorphism of R, that is not the identity mapping and that has at least one fixed point p.

q: the order of f, i.e., the smallest number q such that the q-th iterate of f is the identity mapping.

n: the number of fixed points of f.

P: the Dirichlet region with center at 0, where we assume $\rho(p') = 0$ for some point p' with $\pi(p') = p$. (P consists of all points in D that are hyperbolically closer to 0 than to any other point equivalent to 0.)

∂P: the boundary of P in D.

Our starting point is the

Lemma. *If f exists, then some function f^*, $f^* : D \to D$, that represents f, is a rotation $f^*(z) = h \cdot z$, where h is a primitive q-th root of unity and q is the order of f.*

Proof. Take some neighborhood V, $V \subset U$, of p' where the restriction of π is bijective and define a function φ, $\varphi : V' \to V$, by $\varphi = \pi^{-1} \circ f \circ \pi$ for some $V' \subset V$. Next put $f^* = \rho \circ \varphi \circ \rho^{-1}$ in some neighborhood of 0 and use analytic continuation to extend f^* to D. Similarly f^{-1} has a representation in D that must be the inverse of f^*. Apply the Schwarz lemma to f^* and its inverse and conclude that f^* is a rotation. Finally iterate f^* and observe that there are at least two but only finitely many points that are equivalent to 0 and closest to it. Here we have used that R is not simply connected. Since f^* must interchange those points, q exists and the Lemma follows. $\square$

2. Next we establish that for distinct fixed points there is always more than one geodesic that connects them.

Proposition 1. *If f exists, then $f^*(P) = P$, and further fixed points belong to ∂P. If some such fixed point lies in the interior of a side of ∂P, then $q = 2$ and $f^*(z) = -z$.*

For any two fixed points there are at least two hyperbolic geodesics that connect them. Especially in the latter case where $q = 2$ one gets a closed geodesic connecting those two fixed points.

Proof. Since f^* preserves hyperbolic distances, it follows from the definition of P, that $f^*(P) = P$. Obviously 0 is the only fixed point in P. Any point interior to a side of ∂P has only one other point on ∂P equivalent to it. So $f^* \circ f^*$ is the identity mapping and $h = -1$.

Connect 0 with the fixed points on ∂P by segments and thus confirm the remaining statements. $\square$

Proposition 2. *If $g = 0$, then $n \leq 2$.*

Proof. If f exists, the Lemma and Proposition 1 are applicable. Choose for $n > 1$ one point on ∂P representing some fixed point and then apply f^* and its iterates to that point thus obtaining the vertices of a regular q-gon. Connect each vertex with 0 by a segment and call all points in P and ∂P between two such adjacent segments a sector. Any two such adjacent segments form a loop that for $g = 0$ divides R. Therefore all points on ∂P in a sector have their equivalent ones lying in the same sector. Since f^* maps each sector onto another one, no additional fixed point exists. $\square$

Remark 1. That Proposition 2 also holds in the remaining cases where R is either simply connected or is not hyperbolic, is easily confirmed.

3. Finally we shall prove

Proposition 3. *If* $n > 2$, *then* $(q-1) \cdot (n-2) \leq 2g$.

Proof. Choose a point on ∂P representing some fixed point and construct q sectors as in the proof of Proposition 2. Since in each sector lies a point representing another fixed point, the q sectors are connected in R. If $n = 3$, the construction is complete. Otherwise sweep out a sector by moving one segment positively into the other one. List one representative of each additional fixed point in the order they thus occur as $1, 2, \ldots, n-2$. Do not use the last fixed point and keep only the points $1, 2, \ldots, n-3$. Observe that the segments from 0 to the points $1, 2, \ldots, n-3$ define subsectors labelled as $1', 2', \ldots, (n-2)'$ in that same order. Apply f^* and its iterates to those $n-3$ points and their segments and get in each sector in a similar way subsectors. Our main claim is that all segments together do not subdivide R. Note that in R we get around each point $1, 2, \ldots, n-3$ exactly q sectors.

Take some first subsector $1'$ and the segment from 0 to 1 on its boundary. This sector lies to the right when one goes from 0 to 1. Arriving at 1 we will have the same subdomain to our right coming back along some segment from 1 to 0 but in some other subdivided sector. Therefore some subsector labelled $2'$ lies to the right and is in the same subdomain as subsector $1'$. Thus continuing conclude that each subsector $2'$ is connected with some subsector $3'$ etc. Since all subsectors labelled $(n-2)'$ contain the last not used fixed point, they are connected. Therefore each subsector is connected with some subsector labelled $(n-2)'$ and our claim is correct. Note also that the set S of all segments is connected.

The set $R - S$ is either planar or not. In the latter case make a cut along a loop that starts at p and does not divide $R - S$. Cut if possible the remaining domain along another similarly defined loop. Iterate this process as necessary in order to dissect $R - S$ into some domain R^* that is planar. Let d be the number of loops needed and let T be the set of all loops. If R is compact, then R^* is simply connected because its boundary is connected. Otherwise R^* is homeomorphic to a planar domain Z, where we can choose the boundary corresponding to $S \cup T$ as the outer boundary that separates Z from ∞. The inner boundary of Z represents the boundary of R. Finally connect in P the non-used fixed point with 0 by q segments and thus cut R^* into exactly q pieces as it is readily seen in Z. Here we suppose that T and those segments intersect at p only.

Now apply the Euler-Poincaré formula $2g - 2 = -V + E - F$ to the thus cut-up R. Here the boundary of R is irrelevant and can be ignored. Obviously $V = n$ and $F = q$. Counting all segments as well as all loops, we obtain $E = (n - 1) \cdot q + d$. Hence $(q - 1) \cdot (n - 2) + d = 2g$ holds. $\square$

Corollary. $n \leq 2g + 2$ and equality holds for $g > 0$ only if $q = 2$.

  If $q > 2$, then $n \leq g + 2$ where equality holds for $g > 0$ only if $q = 3$.

  If $n > 2$, then $q \leq 2g + 1$ where equality holds only if $n = 3$.

  If $n > 3$, then $q \leq g + 1$ where equality holds only if $n = 4$.

Remark 2. Our method does not cover the case of the torus. However if one fixed point is removed, we can proceed as above because a punctured torus is hyperbolic. Hence if $q > 2$, $n \leq 3$ by the Corollary and therefore the number of fixed points does not exceed 4.

In the case $q = 2$ we proceed similarly as in the proof of Proposition 1. Using uniformization of the torus, one obtains a parallelogram as a normal form. Choose its center at 0 and assume it is a fixed point. The corresponding function $f^*(z) = -z$ maps the parallelogram onto itself. Here only the midpoints of the sides and the vertices are additional fixed points. Therefore the total number does not exceed 4.

References

[F]   H.M. Farkas. Remarks on automorphisms of compact Riemann surfaces. Annals of Mathematics Studies 79, p. 121–144. Princeton, New Jersey, 1974.

[M]   C.D. Minda. Fixed points of analytic self-mappings of Riemann surfaces. Manusc. Math. 27 (1979), 391–399.

[P/L]   E. Peschl and M. Lehtinen. A conformal self-map which fixes three points is the identity. Ann. Acad. Sc. Fenn. Ser A. I 4 (1978/79), 85–86.

[S]   N. Suita. On fixed points of conformal self-mappings. Hokkaido Math. J. 10 (1981) Sp., 667–671.

Department of Mathematics
Kent State University
Kent, OH 44242, U.S.A.

Complex
Analysis

Edited by
J. Hersch and A. Huber

Birkhäuser Verlag
Basel 1988

H. L. Royden

The Variation of Harmonic Differentials and their Periods

The purpose of the present paper is to construct harmonic and holomorphic differentials with suitably prescribed periods on a compact Riemann surface of genus g. We use these constructions to investigate the variation of the period matrices along a curve $W_{t\mu}$ in Teichmüller space given by a linear family $t\mu$ of Beltrami differentials. In the case of a Teichmüller geodesic, i.e., when $t\mu$ is a Teichmüller differential $t\overline{Q}/|Q|$, we obtain some convexity properties for the real period matrix. The formulae we derive for the variation of the harmonic differential with given periods and for the period matrix are exact. We thus have the variations of all orders, and it is by looking at the second variation that we obtain our convexity result.

§1 Harmonic Differentials and the Period Matrix

Let W be a Riemann surface. A differential 1-form (or differential) $\alpha = pdx + qdy$ on W is said to be exact if there is a function φ on W with $\alpha = d\varphi = \varphi_x dx + \varphi_y dy$. The form α is said to be closed if $d\alpha = (q_x - p_y)dxdy$ is zero. If $\alpha = pdx + qdy$ and $\beta = rdx + sdy$ are two 1-forms, we define their wedge product by

$$\alpha \wedge \beta = (ps - qr)dxdy \ .$$

The processes of exterior differentiation and wedge product depend only on the differential structure of W, not on its complex structure. In contrast, the conjugation operation $*$ defined by

$$*\alpha = -qdx + pdy$$

depends on the complex structure, but is invariant under a holomorphic change of coordinate. Note that $* * \alpha = -\alpha$. We say that a form β is co-closed if $d * \beta = 0$ and co-exact if $\beta = *d\varphi$ for some function φ. A form ω which is both closed and co-closed is said to be harmonic.

We shall write $\alpha * \beta$ for $\alpha \wedge (*\beta)$. Then $\alpha * \beta = \beta * \alpha$, and we get a positive-definite inner product on the space of 1-forms by setting

$$(\alpha, \beta) = \int_W \alpha * \beta \, .$$

The forms α with $\|\alpha\|^2 = (\alpha, \alpha) < \infty$ are said to be square integrable. The space of square-integrable forms is a Hilbert space H, and H has the orthogonal decomposition

$$H = \mathcal{E} \oplus \mathcal{E}^* \oplus \mathcal{H} \, ,$$

where $\mathcal{E}$ is the space of exact forms, $\mathcal{E}^*$ the space of co-exact forms and $\mathcal{H}$ the space of harmonic forms (See [1], [3]). The forms in $\mathcal{H}$ are of class C^∞. If α is a smooth 1-form, then

$$\alpha = d\varphi + *d\psi + \omega \, ,$$

where $d\varphi$ and $*d\psi$ have the same smoothness as α and ω is harmonic.

Let W have a canonical homology basis $\{A_j, B_j\}$, where the A_j and B_j are smooth simple closed curves with Kronecker intersection numbers given by

$$[A_j \times B_k] = \delta_{jk}$$
$$[A_j \times A_k] = 0$$
$$[B_j \times B_k] = 0 \, .$$

When we are not interested in the intersection properties, we often set $C_{j+g} = B_j, 1 \leq j \leq g$.

By a 1-cycle C on W we mean a formal linear combination of simple closed curves. Then the period

$$\int_C \alpha$$

of a closed differential α around C depends only on the homology class of C. Since $\{C_j\}$ forms a basis for the homology of W, the periods of α around the curves C_j determine its periods completely. A closed form α is exact if and only if it has zero periods around each closed curve.

By the orthogonal decomposition of H each closed form α may be written

$$\alpha = d\varphi + \omega \ ,$$

where ω is harmonic. Since it is easy to construct a closed form with prescribed periods around the cycles $\{A_j, B_j\}$, orthogonal projection gives the existence of a harmonic form ω with prescribed periods. If E is orthogonal projection onto the space of exact forms and α a closed differential with the given periods, then

$$\omega = \alpha - E\alpha$$

is the harmonic form with these periods.

A harmonic form with zero periods is exact and therefore identically zero. Thus a harmonic form is uniquely determined by its periods on the homology basis $\{A_j, B_j\}$. Let ω_j be the harmonic differential whose period around C_k is δ_{jk}. Since we can construct a harmonic form with arbitrarily prescribed periods by taking a suitable linear combination of the ω_j, the forms ω_j constitute a basis for the harmonic differentials on W. Thus the real dimension of the space of harmonic forms is $2g$.

Let $M = [m_{jk}]$ be the symmetric matrix given by

$$m_{jk} = \int_W \omega_j * \omega_k \ .$$

Then M represents the inner product on the space $\mathcal{H}$ in the sense that $m_{jk} = (\omega_j, \omega_k)$ and

$$\left(\xi^j \omega_j, \ \eta^k \omega_k \right) = \xi^T M \eta \ .$$

Consequently, M is positive definite. We call M the real period matrix of W, for reasons which will be apparent later.

Let α and α' be two closed forms with periods a_j, b_j and a_j', b_j', respectively, around $\{A_j, B_j\}$. The Riemann bilinear relation asserts

$$\int_W \alpha \wedge \alpha' = \sum_j \left(a_j b_j' - a_j' b_j \right) \ .$$

Define $(g \times g)$ matrices $A, B,$ and C by setting

$$a_{jk} = (\omega_j, w_k)$$
$$c_{jk} = (w_{j+g}, w_k) = (w_k, w_{j+g})$$
$$b_{jk} = (w_{j+g}, w_{k+g}) \ .$$

The matrices A and B are symmetric and positive definite, and the period matrix M has the block form

$$M = \begin{bmatrix} A & C^T \\ C & B \end{bmatrix} \,.$$

Since $a_{jk} = \int \omega_j \wedge (*\omega_k)$, the Riemann bilinear relation tells us that a_{jk} is the period of $*\omega_k$ around B_j. Similarly $-c_{jk}$ is the period of $*\omega_k$ around A_j, c_{jk} is the period of $*\omega_{k+g}$ around A_j, and c_{kj} is the period of $*\omega_{k+g}$ around B_j. Let

$$\omega = \sum_{}^{2g} \xi^k \omega_k \,,$$

and let c_j be the period $*\omega$ around C_j. Then

$$\begin{bmatrix} c_1 \\ \vdots \\ c_{2g} \end{bmatrix} = \begin{bmatrix} -C & -B \\ A & C^T \end{bmatrix} \begin{bmatrix} \xi^1 \\ \vdots \\ \xi^{2g} \end{bmatrix}$$

We observe that $*\omega = \sum c_j \omega_j$. Hence the matrix

$$S = \begin{bmatrix} -C & -B \\ A & C^T \end{bmatrix}$$

not only gives the periods of $*\omega_k$, but also represents the $*$ operator on the space $\mathcal{H}$. Since $** = -1$, we have $S^2 = -I$, and so

$$\begin{aligned} BA = I + C^2 && CB = BC^T \\ AC = C^T A && AB = I + (C^T)^2 \,. \end{aligned}$$

§2   Abelian Differentials and their Periods

So far we have considered real differential forms. In this section we consider complex valued differential forms. A complex valued form α is said to be *pure* if $*\alpha = -i\alpha$. Thus a form α is pure if and only if it has an expression

$$\alpha = A\,dz$$

in terms of a holomorphic coordinate z. A form β is said to be *anti-pure* if $*\beta = i\beta$. These are the forms

$$\beta = B\,\overline{dz} \,.$$

For any form α, the form $\alpha + i * \alpha$ is pure and $\alpha - i * \alpha$ anti-pure, and so α is the unique sum

$$\alpha = \frac{1}{2}(\alpha + i * \alpha) + \frac{1}{2}(\alpha - i * \alpha)$$

of a pure and an anti-pure differential. If $\alpha = A dz + B d\bar{z}$, then we have

$$d\alpha = \frac{1}{2}\left(\frac{\partial B}{\partial z} - \frac{\partial A}{\partial \bar{z}}\right) dz d\bar{z} = -i\left(\frac{\partial B}{\partial z} - \frac{\partial A}{\partial \bar{z}}\right) dx dy$$

If a pure form $\alpha = A dz$ is closed, then $\frac{\partial A}{\partial \bar{z}} = 0$, and A is holomorphic. In this case we call α a holomorphic differential or an Abelian differential. Note that a closed pure form is also co-closed and hence harmonic. The closed anti-pure forms are the anti-holomorphic forms, that is, those whose conjugates are holomorphic.

Every real harmonic form ω is the real part of the Abelian differential $w = \omega + i * \omega$. Since we can prescribe the periods of ω arbitrarily, it follows that we can prescribe the real parts of an Abelian differential and this determines the differential uniquely. A normalization by prescribing the real parts of the periods is called a *real* normalization. Since there are $2g$ periods, the dimension of the Abelian differentials over the reals is $2g$. Hence the dimension of this space over the complex field is g. Every (complex) harmonic differential ω is the sum of a holomorphic differential $\frac{1}{2}(\omega + i * \omega)$ and an anti-holomorphic one $\frac{1}{2}(\omega - i * \omega)$.

Let w be a holomorphic differential. Then the norm of w is given by

$$\|w\|^2 = \int w * \overline{w} = i \int w \wedge \overline{w}$$
$$= i(a_j \overline{b}_j - \overline{a}_j b_j) \, ,$$

where a_j, b_j are the periods of w around A_j, B_j. From this we see that $w = 0$ if all of its A-periods vanish. This means that the linear map taking each holomorphic form into its set of A-periods is one-to-one. Since the space of holomorphic forms and the space of possible A-periods both have complex dimension g, the map must be onto. Consequently, we can prescribe the A-periods of a holomorphic form arbitrarily, and the holomorphic form with given A-periods is unique. The *complex* normalization of an Abelian form arises when we prescribe its A-periods.

Let w_j be the Abelian differential whose periods around A_k are δ_{jk}. For a given Riemann surface W the periods

$$\pi_{jk} = \int_{B_k} w_j$$

are determined. The matrix $\Pi = [\pi_{jk}]$ is called the *complex* period matrix or Riemann matrix for the Riemann surface W.

The wedge product of two pure differentials α and β is always zero, since we have $A dz \wedge B dz = AB dz \wedge dz$, and $dz \wedge dz = 0$. Thus $w_j \wedge w_k = 0$, and the bilinear relation gives us

$$0 = \int w_j \wedge w_k = \pi_{kj} - \pi_{jk} \ .$$

Thus Π is a symmetric matrix. The matrix $Q = [q_{jk}]$ defined by

$$q_{jk} = \frac{1}{2} \int_W w_j * \overline{w}_k$$

represents the Hermitian inner product on the space of Abelian differentials, and is thus Hermitian and positive definite. Since $\overline{w}_k$ is anti-pure,

$$q_{jk} = \frac{i}{2} \int_W w_j \wedge \overline{w}_k = \frac{1}{2} \left(\overline{\pi}_{kj} - \pi_{jk} \right)$$
$$= \Im \pi_{jk} \ .$$

Thus $Q = \Im \Pi$, and so Π has positive definite imaginary part.

We next express the Abelian differentials w_j in terms of the harmonic differentials ω_k. The period of w_j is δ_{jk} around A_k and is π_{jk} around B_k, and so we have

$$w_j = \omega_j + \sum_k \pi_{jk} \omega_{k+g} \ ,$$

and

$$*w_j = *\omega_j + \sum \pi_{jk} * \omega_{k+g} \ .$$

Since $*w_j = -i w_j$, we have

$$\begin{bmatrix} -C & -B \\ A & C^T \end{bmatrix} \begin{bmatrix} I \\ \Pi \end{bmatrix} = -i \begin{bmatrix} I \\ \Pi \end{bmatrix}$$

Hence $-C - B\Pi = -iI$, or $B\Pi = -C + iI$. From this we get

$$\Pi = -B^{-1}C + iB^{-1} \ .$$

Writing $\Pi = P + iQ$ gives

$$Q = B^{-1}$$

and

$$P = -B^{-1}C \, ,$$

expressing Π in terms of B and C. We can also express A, B, C in terms of P and Q: After a few matrix calculations one gets

$$M = \begin{bmatrix} A & C^T \\ C & B \end{bmatrix} = \begin{bmatrix} Q + PQ^{-1}P & -PQ^{-1} \\ -Q^{-1}P & Q^{-1} \end{bmatrix} \, ,$$

and

$$S = \begin{bmatrix} -C & -B \\ A & C^T \end{bmatrix} = \begin{bmatrix} Q^{-1}P & -Q^{-1} \\ Q + PQ^{-1}P & -PQ^{-1} \end{bmatrix} \, .$$

§3 The Variation of the Harmonic Differentials

Let W_0 denote a surface W with a fixed conformal structure, which we take to be our base structure. A new complex structure on W can be expressed in terms of the base structure by giving a Beltrami differential μ on W_0. This is a form of type $(-1, 1)$ on W_0, i.e., one that varies with conformal changes of coordinate on W_0 so that $\mu \frac{d\bar{z}}{dz}$ remains invariant. We also have

$$\|\mu\| = \sup_{W_0} |\mu| < 1 \, .$$

Holomorphic functions on the Riemann surface W_μ specified by μ are those f which satisfy the complex Beltrami equation

$$\frac{\partial f}{\partial \bar{z}} = \mu \frac{\partial f}{\partial z} \, .$$

This is equivalent to saying that the differentials of the form

$$A(dz + \mu d\bar{z})$$

are pure on W_μ.

If we let $\star$ denote the conjugation operator $\star(\mu)$ on W_μ, then

$$\star(dz + \mu\overline{dz}) = -i(dz + \mu\overline{dz})$$
$$\star(\overline{dz} + \overline{\mu}dz) = i(\overline{dz} + \overline{\mu}dz) \, .$$

Subtracting μ times the second equation from the first gives

$$(1 - |\mu|^2) \star dz = -i \left[(1 + |\mu|^2) \, dz + 2\mu d\bar{z} \right]$$

Equating real and imaginary parts, we get

$$\star dx = (1 - |\mu|^2)^{-1} \left[(1 + |\mu|^2 - 2\sigma)\, dy + 2\tau dx \right]$$
$$\star dy = -(1 - |\mu|^2)^{-1} \left[(1 + |\mu|^2 + 2\sigma)\, dx + 2\tau dy \right] \;,$$

where $\mu = \sigma + i\tau$.

Setting $\alpha = pdx + qdy$ and $\star\alpha = \widetilde{p}dx + \widetilde{q}dy$ gives

$$\begin{bmatrix} \widetilde{p} \\ \widetilde{q} \end{bmatrix} = \frac{1}{1 - |\mu|^2} \begin{bmatrix} 2\tau & -(1 + |\mu|^2 - 2\sigma) \\ (1 + |\mu|^2 + 2\sigma) & -2\tau \end{bmatrix} \begin{bmatrix} p \\ q \end{bmatrix} \;,$$

and

$$\alpha \star \alpha = (p\widetilde{q} - q\widetilde{p})dxdy$$
$$= [p, q]G \begin{bmatrix} p \\ q \end{bmatrix} dxdy \;,$$

where

$$G = \frac{1}{1 - |\mu|^2} \begin{bmatrix} 1 + |\mu|^2 - 2\sigma & -2\tau \\ -2\tau & 1 + |\mu|^2 + 2\sigma \end{bmatrix} \cdot$$

Harmonic differentials on W_μ with prescribed periods can be constructed using the orthogonal projection F onto the exact forms with respect to the new inner product

$$(\alpha; \beta) = \int \alpha \star \beta = (\alpha, G\beta) \;.$$

If α is a closed differential with the desired periods, then

$$\alpha - F\alpha$$

will be harmonic (in the conformal structure of W_μ) and have the same periods as α.

Let E be the orthogonal projection onto the exact forms, using the inner product

$$(\alpha, \beta) = \int \alpha * \beta$$

on W_0. We give a formula for F in terms of E and G.

Let $T^*, T^\star$ be the adjoints of an operator T with respect to the inner products (α, β) and $(\alpha; \beta)$, respectively. Then

$$
\begin{aligned}
(\alpha; T^\star \beta) = (T\alpha; \beta) &= (T\alpha, G\beta) \\
&= (\alpha, T^* G\beta) \\
&= (\alpha, GG^{-1}T^* G\beta) \\
&= (\alpha; G^{-1}T^* G\beta) \, .
\end{aligned}
$$

Thus $T^\star = G^{-1}T^* G$.

Observe that $T^\star$ remains the same if we replace G by $c^{-1}G$ for some positive constant c. By a suitable choice of c, we can write $G = c(I - \Gamma)$ where $\|\Gamma\| < 1$, and $\Gamma = \Gamma^*$. Then

$$
T^\star = (1 - \Gamma)^{-1}T^*(I - \Gamma) \, .
$$

The orthogonal projection F is uniquely characterized by

$$
\begin{aligned}
F^2 = F, \quad F = F^\star \\
FE = E, \quad EF = F \, .
\end{aligned}
$$

These equations are seen to be satisfied by the operator

$$
\begin{aligned}
F = E(I - E\Gamma E)^{-1}(I - \Gamma) \\
= (I - E\Gamma)^{-1}E(I - \Gamma) \, .
\end{aligned}
$$

Hence

$$
\begin{aligned}
I - F = I - (I - E\Gamma)^{-1}E(I - \Gamma) \\
= (I - E\Gamma)^{-1}\left[I - E\Gamma - E(I - \Gamma)\right] \\
= (I - E\Gamma)^{-1}(I - E) \, .
\end{aligned}
$$

If ω_j and $\widetilde{\omega}_j$ are the harmonic differentials on W_0 and W_μ, respectively, with periods δ_{jk} around C_k, then

$$
\widetilde{\omega}_j = \omega_j - F\omega_j = (I - F)\omega_j \, .
$$

Since $E\omega_j = 0$, we have

$$
\begin{aligned}
\widetilde{\omega}_j &= (I - E\Gamma)^{-1}\omega_j \\
&= \omega_j + E\Gamma\omega_j + E\Gamma E\Gamma\omega_j + E\Gamma E\Gamma E\Gamma\omega_j + \dots \, .
\end{aligned}
$$

This is an exact formula for $\widetilde{\omega}_j$.

Although we have been considering formulae for the projection F as an operator on the Hilbert space of square integrable differentials, we note that the projection operator E is also a bounded operator on the Banach space $C^{k+\alpha}$ of differentials whose k-th derivative satisfy a Hölder condition with exponent α. Thus the formula for $\widetilde{\omega}_j$ converges in these spaces if Γ is small enough, i.e., if μ is sufficiently small.

§4 Variation of the Period Matrix

Let $\widetilde{M}$ be the "real" period matrix on W_μ. Thus $\widetilde{M} = [\widetilde{m}_{jk}]$ where

$$\widetilde{m}_{jk} = \int \widetilde{\omega}_j \star \widetilde{\omega}_k = (\widetilde{\omega}_j;\ \widetilde{\omega}_k) \ .$$

Since $\star\widetilde{\omega}_k$ is closed and $\widetilde{\omega}_j - \omega_j$ is exact, we have

$$\begin{aligned}
\widetilde{m}_{jk} &= \int \omega_j \star \widetilde{\omega}_k = (\omega_j;\ \widetilde{\omega}_k) \\
&= (\omega_j,\ G\widetilde{\omega}_k) \\
&= c(\omega_j,\ (1 - \Gamma)\widetilde{\omega}_k) \ ,
\end{aligned}$$

where

$$G = c(I - \Gamma) \ .$$

The formula

$$\widetilde{\omega}_k = (I - E\Gamma)^{-1}\omega_k$$

of the last section gives

$$\begin{aligned}
(I - \Gamma)\widetilde{\omega}_k &= (1 - \Gamma)(1 - E\Gamma)^{-1}\omega_k \\
&= (I - E\Gamma + E\Gamma - \Gamma)(1 - E\Gamma)^{-1}\omega_k \\
&= \omega_k - (I - E)\Gamma(I - E\Gamma)^{-1}\omega_k \ .
\end{aligned}$$

Hence

$$\widetilde{m}_{jk} = c(\omega_j, \omega_k) - c(\omega_j, (I - E)\Gamma(1 - E\Gamma)^{-1}\omega_k) \ .$$

Since $E^* = E$ and $E\omega_j = 0$, we have

$$\begin{aligned}
\widetilde{m}_{jk} &= c[(\omega_j, \omega_k) - c(\omega_j, \Gamma(I - E\Gamma^{-1})\omega_k] \\
&= c\,[(\omega_j, \omega_k) - (\omega_j, \Gamma\omega_k) - (\omega_j, \Gamma E\Gamma\omega_k) - (\omega_j, \Gamma E\Gamma E\Gamma\omega_k) - \ldots] \ .
\end{aligned}$$

If we take $c = 1$,

$$\Gamma = I - G = \frac{2}{1 - |\mu|^2} \begin{bmatrix} \sigma - |\mu|^2 & \tau \\ \tau & -\sigma - |\mu|^2 \end{bmatrix} .$$

and

$$\begin{aligned} \widetilde{m}_{jk} &= m_{jk} - (\omega_j, \Gamma(I - E\Gamma)^{-1}\omega_k) \\ &= m_{jk} - (\omega_j, \Gamma\omega_k) - (\omega_j, \Gamma E\Gamma\omega_k) - (\omega_j, \Gamma E\Gamma E\Gamma\omega_k) - \ldots . \end{aligned}$$

Successive terms are $O(|\mu|^n)$, but because Γ is not linear in μ, the n-th term is not homogeneous in $|\mu|$.

More useful expansions can be obtained when $|\mu|$ is constant on W_0. This special case includes, of course, the Teichmüller geodesics,

$$\mu = t\frac{\overline{Q}}{|Q|}$$

with Q a holomorphic quadratic differential and $|t| < 1$. We express Beltrami differentials of constant absolute value by writing them as $t\mu$, where $|t| < 1$ and $|\mu| \equiv 1$. For such a μ we have

$$G = \frac{1 + t^2}{1 - t^2} \left(I - \frac{2t}{1 + t^2}\Gamma \right) ,$$

where

$$\Gamma = \begin{bmatrix} \sigma & \tau \\ \tau & -\sigma \end{bmatrix} ,$$

and so

$$\begin{aligned} \widetilde{m}_{jk} = \frac{1 + t^2}{1 - t^2} \Bigg(m_{jk} &- \frac{2t}{1 + t^2}(\omega_j, \Gamma\omega_k) - \left(\frac{2t}{1 + t^2}\right)^2 (\omega_j, \Gamma E\Gamma\omega_k) \\ &- \left(\frac{2t}{1 + t^2}\right)^3 (\omega_j, \Gamma E\Gamma E\Gamma\omega_k) - \ldots \Bigg) . \end{aligned}$$

If we let $u = 2t/(1 + t^2)$, we get

$$c = \frac{1 + t^2}{1 - t^2} = (1 - u^2)^{-1/2} .$$

Writing M_n for the matrix whose elements are $(\omega_j, (\Gamma E)^{n-1}\Gamma\omega_k)$, we have

$$\widetilde{M} = (1 - u^2)^{-1/2} \left(M - uM_1 - u^2 M_2 - \ldots - u^n M_n - \ldots \right) .$$

Let us consider the family of Riemann surfaces $W_{t\mu}$ as a curve in Teichmüller space (or more generally in the space of all complex structures on W without regard to equivalence). If μ is a Teichmüller differential $\overline{Q}/|Q|$, then this curve is a geodesic in Teichmüller space. A natural parameter (when $|\mu| \equiv 1$) is

$$s = \frac{1}{2} \log \frac{1+t}{1-t} \, ,$$

so that $t = \tanh s$, and $u = \tanh 2s$. In the case of Teichmüller geodesics the parameter s is the Teichmüller arc length along the geodesic. In terms of s the formula for the period matrix becomes

$$M(s) = \widetilde{M} = \cosh 2s (M - M_1 \sinh s - M_2 (\sinh 2s)^2 - \ldots) \, .$$

Observe that the matrix

$$M_2 = [(\omega_j, \Gamma E \Gamma \omega_k)] = [(E \Gamma \omega_j, E \Gamma \omega_k)]$$

is positive semi-definite. If we take the second derivative of M with respect to s we obtain[1]

$$M''(0) = 4M(0) - 8M_2 \leq 4M(0) \, .$$

Since we may take any point on our geodesic (or curve with $|\mu| \equiv 1$) in place of the origin and carry out the same calculations, we must have

$$M''(s) \leq 4M(s) \, .$$

If we set $M = (\cosh 2s)X$, then

$$(\cosh 2s)X'' + 4(\sinh 2s)X' \leq 0 \, ,$$

and so

$$\left[(\cosh 2s)^2 X' \right]' \leq 0 \, .$$

In terms of the parameter $u = \tanh 2s$, we have $\dfrac{du}{ds} = 2(\cosh 2s)^{-2}$, or

$$\frac{ds}{du} = \frac{1}{2}(\cosh 2s)^2 \, .$$

Hence

$$\frac{d}{du}\left(\frac{d}{du}X\right) \leq 0 \, ,$$

---

[1]   We write $A \leq B$ for real symmetric matrices to mean that $B - A$ is positive semi-definite.

and we see that the matrix-valued function X is concave as a function of the parameter u. We summarize in the following proposition:

Proposition: *Let μ be a Beltrami differential on W_0 with $|\mu| \equiv 1$. Along the curve $W_{t\mu}$ the matrix*

$$X = \left[(\cosh 2s)^{-1} M(s)\right] = \left[(1 - u^2)^{1/2} M\right]$$

is a concave function of the parameter $u = 2t/(1 + t^2) = \tanh 2s$. Here M is the real period matrix of $W_{t\mu}$.

When μ is a Teichmüller differential on a torus, the matrix X is a linear function of u. More generally, let

$$\mu = \frac{\omega + i * \omega}{\omega - i * \omega}$$

for any Abelian differential $\omega + i * \omega$ with $\omega = \xi^j \omega_j$ and $*\omega = \eta^j \omega_j$. Then the (2×2) matrix

$$\begin{bmatrix} \xi^T X \xi & \xi^T X \eta \\ \eta^T X \xi & \eta^T X \eta \end{bmatrix}$$

is linear in the parameter u.

Bibliography

[1] Ahlfors, L. and Sario, L., *Riemann Surfaces*. Princeton University Press, Princeton, NJ, 1960.

[2] Pfluger, A. *Theorie der Riemannschen Flächen*. Springer, Berlin-Göttingen-Heidelberg, 1957.

[3] Royden, H. L., "Function theory on compact Riemann surfaces", *J. d'Analyse Math.* **18** (1967), pp. 295–327.

This research sponsored by NSF grant DMS–8603148

Department of Mathematics
Stanford University
Stanford, California 94305
USA

Kurt Strebel

On the Extremality and Unique Extremality of Certain Teichmüller Mappings

Introduction

1. A quasiconformal mapping f of a domain G onto a domain G' with maximal dilatation K is called extremal, if every qc mapping $\tilde{f}$ which agrees with f on the boundary of G and is homotopic to f has a maximal dilatation $\tilde{K} \geq K$. It is called uniquely extremal, if the strict inequality $\tilde{K} > K$ holds whenever $\tilde{f} \neq f$. Since the maximal dilatations of a qc mapping f and of its inverse f^{-1} are the same, the mapping f^{-1} is extremal (uniquely extremal) if and only if f is.

A Teichmüller mapping is a qc mapping f with a complex dilatation of the form $\kappa = k\frac{\bar{\phi}}{|\phi|}$; where k is a constant, $0 < k < 1$, and ϕ is a holomorphic quadratic differential.

If ϕ has finite norm $\| \phi \| = \iint |\phi| dx dy < \infty$, the Teichmüller mapping associated with ϕ is uniquely extremal, for each k, $0 < k < 1$ ([4], [2], [5]). If ϕ has infinite norm, the mapping f can be uniquely extremal, extremal but not unique, and not extremal. Examples are, respectively, the affine mappings of a parallel strip, of a chimney region or an angular region, with an angle α different from π and 2π ([3]).

In a recent paper [7] of the author, the example of the parallel strip has been modified in the following way. The domain G is a vertical parallel strip with a distinguished closed set E of boundary points containing the two points at infinity of the strip. The strip G is mapped onto a strip G' by a horizontal stretching $f : z = x + iy \rightarrow Kx + iy$, $K > 1$. The mappings $\tilde{f}$ of G onto G' agree with f on E, whereas the complementary intervals of E on ∂G are considered as free boundary intervals. Then, f is extremal in this class if and only if at least one of the points $\pm i\infty$ is an accumulation point of E, and it is uniquely extremal if and only if both points are accumulation points of E.

The inverse mapping f^{-1} corresponds to a stretching *along* the strip, and the same statement holds.

2. The situation is quite similar, if we consider the $K - qc$ mapping

$$w = f(z) \ : \ |w| = |z|^{1/K}, \ \arg w = \arg z$$

of the extended plane onto itself. It is a Teichmüller mapping associated with the quadratic differential

$$\phi(z)dz^2 = \frac{-1}{z^2}dz^2.$$

The trajectories of ϕ are the circles $|z| = \text{const}$. The mapping f is a contraction along the radii $\arg z = \text{const.}$, or rather a stretching along the trajectories of ϕ. The inverse mapping f^{-1} takes w into z, $|z| = |w|^K$, $\arg z = \arg w$; but we rather work with f itself.

Let E be a closed set containing 0 and ∞ as only (possible) accumulation points. In case both 0 and ∞ are isolated points of E, the set E is finite.

The mapping f takes E onto $E' = f(E)$. It is considered as a mapping of $G = \hat{\mathbb{C}} \setminus E$ onto $G' = \hat{\mathbb{C}} \setminus E'$. Let now $\tilde{f}$ be a qc mapping of G onto G' which is equal to f on $\partial G = E$ and homotopic to f. Extremality and unique extremality are meant in this class of mappings.

Theorem 1. *In the above setting, the mapping f is extremal if and only if at least one of the points zero and infinity is an accumulation point of E; f is uniquely extremal if and only if both points zero and infinity are accumulation points of E.*

It should be noticed that there is no question of density of the set E near zero and infinity.

3. The next step is a generalization to Riemann surfaces and the admittance of first order poles. Let ϕ be a rational quadratic differential on a compact Riemann surface R, with at least one second order pole and no poles of order higher than two. Furthermore, assume that the trajectories of ϕ in the neighborhoods of the second order poles P_ν are closed. This amounts to saying that the leading coefficients of the Laurent developments at these poles are negative.

Let f be the Teichmüller mapping of R, with dilatation K and associated with ϕ, onto a Riemann surface R'.

Next, let E be a set of points on R, containing all the poles of ϕ and with no other accumulation points than (possibly) the points P_ν. The set E is mapped onto a set E' by f.

We now consider tha family of all quasiconformal mappings $\tilde{f}$ of $G = R \setminus E$ onto $G' = R' \setminus E'$ which agree with f on E and are homotopic to f. We have

Theorem 2. *The Teichmüller mapping f is extremal in the above class if and only if at least one of the second order poles P_ν is an accumulation point of E; it is uniquely extremal if and only if all P_ν are accumulation points of E.*

Again, there is no question of density of the set E near the points P_ν. It is of course possible that there is only one second order pole and ϕ is holomorphic elsewhere. If ϕ has only one second order pole, the associated mapping is either uniquely extremal or not extremal, depending on whether E is infinite or finite.

Extremality

4. Consider the special case (section 2) and assume that there is a sequence of points $z_n \in E$ tending to infinity. Let $\tilde{f}$ with maximal dilatation $\tilde{K}$ be in our class. The circular dilatation (see [1], pg 105) of $\tilde{f}$ is finite at each point z. At $z = \infty$, this means that

$$H(\infty) = \lim_{r \to \infty} \sup \frac{\max_{|z|=r} |\tilde{f}(z)|}{\min_{|z|=r} |\tilde{f}(z)|} \leq e^{\pi \tilde{K}} \; :$$

Therefore there exist numbers r_0 and σ_0 such that for $r > r_0$ the variation of $\log |\tilde{f}(z)|$ on the circle $|z| = r$ is

$$\sigma(r) = \max_{|z|=r} \log |\tilde{f}(z)| - \min_{|z|=r} \log |\tilde{f}(z)| < \sigma_0. \tag{1}$$

On the other hand, the image $\tilde{\gamma}_n$ of the circle $\gamma_n : |z| = r_n$ contains the point w_n, $|w_n| = |z_n|^{1/K} = r_n^{1/K}$. For all $w \in \tilde{\gamma}_n$ we therefore have the inequality

$$\frac{1}{K} \log r_n - \sigma_0 \leq \log |w| \leq \frac{1}{K} \log r_n + \sigma_0, \quad w \in \tilde{\gamma}_n. \tag{2}$$

Let $\tilde{M}_{m,n}$ be the modulus of the ring domain $\tilde{R}_{m,n}$, bounded by $\tilde{\gamma}_m$ and $\tilde{\gamma}_n$, $m < n$, respectively. It satisfies the double inequality

$$\frac{1}{\tilde{K}} \cdot \frac{1}{2\pi} \{\log r_n - \log r_m\} \leq \tilde{M}_{m,n} \leq \frac{1}{K} \cdot \frac{1}{2\pi} \{\log r_n - \log r_m + 2\sigma_0\}, \quad (3)$$

hence

$$\frac{1}{\tilde{K}} \leq \frac{1}{K} \cdot \left\{1 + \frac{2\sigma_0 \cdot}{\log r_n - \log r_m}\right\}. \quad (4)$$

Letting $n \to \infty$, we find $\tilde{K} \geq K$.

We conclude that every quasiconformal mapping $\tilde{f}$ which agrees with f on the points z_n in a neighborhood of ∞ and lies in the same homotopy class has a maximal dilatation $\tilde{K} \geq K$: the point ∞ is what is called an essential boundary point.

The situation is the same at the point $z = 0$. We therefore conclude that if at least one of the two points 0 and ∞ is an accumulation point of E, f is extremal.

If neither zero nor infinity is an accumulation point of E, we have only finitely many points left. For $|z| \geq r_0$, with sufficiently large r_0, we can replace f by the conformal mapping

$$z \to r_0^{\frac{1}{K}-1} \cdot z. \quad (5)$$

The same is possible near $z = 0$. The new mapping has the same maximal dilatation as f, namely K. By Teichmüller's theorem, there is only one extremal mapping in the homotopy class of f which agrees with f on E. It evidently cannot be the above mapping and therefore has smaller maximal dilatation : f is not extremal. This proves the first part of our theorem in the special case.

5. In the *general case*, let ϕ be the quadratic differential (section 3) and let P_ν be one of its second order poles. The Laurent series, in terms of an arbitrary parameter z, $z = 0 \leftrightarrow P_\nu$, is of the form

$$\Phi(z)dz^2 = \left\{\frac{a_{-2}}{z^2} + \frac{a_{-1}}{z} + a_0 + a_1 z + \ldots\right\} dz^2, \quad a_{-2} < 0.$$

It is possible to introduce a new parameter ζ in a sufficiently small neighborhood of P_ν, such that the quadratic differential is represented by (see [6], pg 29)

$$\phi(z)dz^2 = \frac{a_{-2}}{\zeta^2}d\zeta^2, \quad \zeta \leftrightarrow z.$$

The representation is the same, whether we use it near $\zeta = 0$ or near $\zeta = \infty$, since the transformation $\zeta \to \frac{1}{\zeta}$ evidently does not change it.

The Teichmüller mapping f induces a quadratic differential ψ on the image surface. Again, near $P'_\nu = f(P_\nu)$ we can introduce the natural parameter. The representation of ψ in terms of this parameter is the same as above, and the mapping f is as before in the special case near infinity. Again, P_ν is an essential boundary point if it is an accumulation point of E. Therefore, f is extremal if E accumulates to at least one of the second order poles.

If all the second order poles P_ν are isolated points of E, we can replace f by conformal mappings near the points P_ν, as above in the special case, using the natural parameters. We get a variation $\tilde{f}$ of f which agrees with f on E and lies in the same homotopy class, but is conformal near the points P_ν. Since the extremal problem has a unique Teichmüller solution, the extremal qc mapping has a dilatation smaller than K, hence f is not extremal.

Unique extremality

6. The proof of the unique extremality needs a sharper estimate near the second order poles which are accumulation points of E. This sharpening is possible whenever f is extremal.

Again we start with the special case of the plane and the quadratic differential $\frac{-1}{z^2}dz^2$. Let z_n be a sequence of points of E which tends monotonically to infinity. Assume that $\tilde{f}$ is extremal, i.e. $\tilde{K} = K$. Let $w = \tilde{f}(z)$, $dw = p\,dz + q\,d\bar{z}$. The logarithmic length of the image curve $\tilde{\gamma}_\rho$ of the circle $\gamma_\rho : |z| = \rho$ satisfies

$$2\pi \leq L(\rho) \equiv \int\limits_{\tilde{\gamma}_\rho} |d\log w| = \int\limits_{\tilde{\gamma}_\rho} \frac{|dw|}{|w|} = \int\limits_{\gamma_\rho} \frac{|p\,dz + q\,d\bar{z}|}{|w|}. \tag{6}$$

Integrating from r_0 to r gives, with $z = \rho \cdot e^{i\vartheta}$

$$2\pi \log\frac{r}{r_0} \leq \int\limits_{r_0}^{r} L(\rho)\frac{d\rho}{\rho} = \int\limits_{r_0}^{r}\int\limits_{\gamma_\rho} \frac{|pe^{i\vartheta} - q\cdot e^{-i\vartheta}|}{|w|\cdot\rho}\rho\,d\rho\,d\vartheta. \tag{7}$$

The Jacobian of $\tilde{f}$ is $J(w/z) = |p|^2 - |q|^2$. We thus get, after squaring and using the Schwarz inequality, in the usual way

$$
\left.
\begin{aligned}
\left(\int_{r_0}^{r} L(\rho) \frac{d\rho}{\rho} \right)^2 &\leq \iint_{r_0 < |z| < r} (|p|^2 - |q|^2) \frac{1}{|w|^2} \, dx \, dy \times \\
&\times \iint_{r_0 < |z| < r} \frac{|1 - \kappa e^{-2i\vartheta}|^2}{1 - |\kappa|^2} \frac{1}{\rho^2} \, dx \, dy.
\end{aligned}
\right\}
\tag{8}
$$

The first factor on the right hand side is the logarithmic area $A(r_0, r)$ of the image by $\tilde{f}$ of the annulus $r_0 < |z| < r$. The second factor is estimated above by

$$
K \cdot 2\pi \cdot \log \frac{r}{r_0}.
\tag{9}
$$

We thus get the inequality

$$
\left(\int_{r_0}^{r} L(\rho) \frac{d\rho}{\rho} \right)^2 \leq A(r_0, r) \cdot K 2\pi \log \frac{r}{r_0}.
\tag{10}
$$

We now set $r_0 = |z_m|$, $r = |z_n|$, $n > m$. The points z_m and z_n belong to E. They are mapped, by $\tilde{f}$, onto points w_m, w_n with $|w_m| = |z_m|^{1/K}$, $|w_n| = |z_n|^{1/K}$. Because of the boundedness of the variation of the logarithm, we get, as in (3)

$$
A(r_m, r_n) \leq \frac{2\pi}{K} \{ \log r_n - \log r_m + 2\sigma_0 \},
\tag{11}
$$

and therefore

$$
\left(\int_{r_m}^{r_n} L(\rho) \frac{d\rho}{\rho} \right)^2 \leq 4\pi^2 \left\{ \log \frac{r_n}{r_m} + 2\sigma_0 \right\} \log \frac{r_n}{r_m}.
\tag{12}
$$

Assume now that for some $\delta_1 > 0$ and all $\rho > r_m$ the inequality $L(\rho) \geq 2\pi + \delta_1$ holds. Then, after dividing by $(\log \frac{r_n}{r_m})^2$, we have

$$
(2\pi + \delta_1)^2 \leq 4\pi^2 \left\{ 1 + \frac{2\sigma_0}{\log \frac{r_m}{r_n}} \right\},
\tag{13}
$$

which becomes contradictory for large enough values of r_n. We now replace δ_1 by $\delta_2 < \delta_1$ and r_m by r_n, r_n by r_p, $p > n$, and repeat the argument, etc.

We get an increasing sequence of radii $\rho_j \to \infty$ with the property that the logarithmic length of the image curve tends to 2π, and hence the variation of the logarithm tends to zero,

$$\tau_j := \log m_j'' - \log m_j' \to 0, \tag{14}$$

with $m_j'' = \max_{|z|=\rho_j} |\tilde{f}(z)|$, $m_j' = \min_{|z|=\rho_j} |\tilde{f}(z)|$.

7. The image γ_j of $|z| = \rho_j$ becomes more and more like a circle. But since there is, in general, no point $z_n \in E$ with $|z_n| = \rho_j$ and hence $w_n \in \gamma_j$, $|w_n| = \rho_j^{1/K}$, we do not know the approximate radius of γ_j.

We choose z_m and z_n in E such that $|z_m| < \rho_j < |z_n|$. The modulus M of the image of the annulus $\rho_j < |z| < r_n$ satisfies

$$\frac{1}{K}\frac{1}{2\pi}\log\frac{r_n}{\rho_j} \le M \le \frac{1}{2\pi}\left\{\log r_n^{1/K} + \sigma_0 - \log m_j'' + \tau_j\right\}. \tag{15}$$

Therefore

$$\log m_j'' \le \frac{1}{K}\log\rho_j + \sigma_0 + \tau_j. \tag{16}$$

We conclude that

$$\overline{\lambda} = \overline{\lim}\left\{m_j'' - \frac{1}{K}\log\rho_j\right\} \le \sigma_0. \tag{17}$$

Similarly, the annulus $|z_m| < |z| < \rho_j$ has an image, by $\tilde{f}$, with modulus M satisfying

$$\frac{1}{K}\cdot\frac{1}{2\pi}\log\frac{\rho_j}{r_m} \le M \le \frac{1}{2\pi}\left\{\log m_j' + \tau_j - \log r_m^{1/K} + \sigma_0\right\}, \tag{18}$$

hence

$$\frac{1}{K}\log\rho_j \le \log m_j' + \sigma_0 + \tau_j. \tag{19}$$

This leads to a lower estimate of the lim.inf.,

$$\underline{\lambda} = \underline{\lim}\left\{\log m_j' - \frac{1}{K}\log\rho_j\right\} \ge -\sigma_0. \tag{20}$$

8. In order to see that $\underline{\lambda} = \overline{\lambda}$, let $\varepsilon > 0$. Choose j such that

$$\log m_j'' - \frac{1}{K}\log\rho_j > \overline{\lambda} - \varepsilon, \tag{21}$$

then $l > j$ such that

$$\log m_l' - \frac{1}{K} \log \rho_l < \underline{\lambda} + \varepsilon. \tag{22}$$

The inequality for moduli gives

$$\frac{1}{K} \frac{1}{2\pi} (\log \rho_l - \log \rho_j) \leq \frac{1}{2\pi} (\log m_l' - \log m_j'' + \tau_l + \tau_j), \tag{23}$$

hence

$$0 \leq \log m_l' - \frac{1}{K} \log \rho_l - \left(\log m_j'' - \frac{1}{K} \log \rho_j \right) + \tau_l + \tau_j < \underline{\lambda} - \overline{\lambda} + \tau_l + \tau_j + 2\varepsilon. \tag{24}$$

As τ_l and τ_j become arbitrarily small and ε is an arbitrary positive number,

$$0 \leq \underline{\lambda} - \overline{\lambda} \leq 0, \tag{25}$$

which proves the convergence. We will actually see that $\lambda = 0$.

9. For this last step we need the Teichmüller module theorem ([8], pg 649). Let $R : r_1 < |z| < r_2$ be an arbitrary annulus, and let R_1 and R_2 be two disjoint ring domains in R, both separating the two boundary circles of R. Denote by M, M_1 and M_2 the moduli of R, R_1 and R_2 respectively. Then, for every $\varepsilon > 0$ there exists a $\delta > 0$ such that if $M_1 + M_2 > M - \delta$, any two points z', z'' in the continuum Γ separating R_1 and R_2 satisfy

$$\log |z'| - \log |z''| < \varepsilon. \tag{26}$$

In other words, the variation of $\log |z|$ on Γ is smaller than ε. The bound δ is independent of M.

Let $\rho_j < |z_n| < \rho_l$. The annulus $R_{jl} : \rho_j < |z| < \rho_l$ is subdivided into the two disjoint annuli $R_{jn} : \rho_j < |z| < r_n$, $R_{nl} : r_n < |z| < \rho_l$, with $r_n = |z_n|$. Let their images, by $\tilde{f}$, be denoted by R_{jl}', R_{jn}' and R_{nl}', with moduli M_{jl}', M_{jn}' and M_{nl}' respectively.
The ring domain R_{jl}' lies in an annulus of the w-plane satisfying

$$\log m_j' \leq \log |w| \leq \log m_l''. \tag{27}$$

Its modulus is, by the result of section 8,

$$\frac{1}{2\pi} \{ \log m_l'' - \log m_j' \} \leq \frac{1}{2\pi} \frac{1}{K} \{ \log \rho_l - \log \rho_j \} + \delta,$$

with δ arbitrarily small for large enough values of j and l. The ring domains R'_{jn} and R'_{nl} are embedded in the annulus (27) and have moduli

$$\frac{1}{K}\,\frac{1}{2\pi}\log\frac{r_n}{\rho_j} \leq M'_{jn}, \qquad \frac{1}{K}\,\frac{1}{2\pi}\log\frac{\rho_l}{r_n} \leq M'_{nl}, \tag{28}$$

and hence

$$M'_{jn} + M'_{nl} \geq \frac{1}{2\pi}\{\log m''_l - \log m'_j\} - \delta. \tag{29}$$

The first term on the right hand side is the modulus of an annulus which contains R'_{jn} and R'_{nl}. Therefore the image of the circle $|z| = r_n$, which separates R'_{jn} from R'_{nl}, has a logarithmic variation which becomes arbitrarily small. Since it contains w_n, $|w_n| = |z_n|^{1/K}$, the difference

$$\log|\tilde{f}(z)| - \frac{1}{K}\log r_n, \qquad |z| = r_n, \tag{30}$$

becomes uniformly arbitrarily small with $n \to \infty$.

We can now apply the same reasoning to an arbitrary circle $|z| = r$, using r_n and r_m, $r_n < r < r_m$, instead of ρ_j and ρ_l. We then get the result that

$$|\log|\tilde{f}(z)| - \frac{1}{K}\log r|, \qquad |z| = r, \tag{31}$$

becomes arbitrarily small with $r \to \infty$, uniformly for all z.

10. The unique extremality in the special case in the plane, if both zero and infinity are accumulation points of E, follows from another application of Teichmüller's module theorem. Let $|z| = r$. Choose ρ_1 close to zero and ρ_2 close to infinity. The image curves of $|z| = \rho_1$ and $|z| = \rho_2$ will be arbitrarily close to the circles $|w| = \rho_i$, $i = 1, 2$, respectively. Therefore the image of $|z| = r$ must be arbitrarily close to a circle, hence a circle for every r.

For the annulus $r_m < |z| < r_n$ the image circles have radii $r_m^{1/K}$ and $r_n^{1/K}$ respectively, and the mapping $\tilde{f}$ is $K-qc$. The only possibility is that $\tilde{f}$ is equal to f in $r_m < |z| < r_n$, up to a rotation. Since $f(z_m) = \tilde{f}(z_m)$, there is no rotation possible, and since m and n are arbitrary, $\tilde{f} = f$.

If only one of the two points zero or infinity is an accumulation point of E, the other one is isolated, and we can replace f by a conformal mapping in a neighborhood of this point, as in section 4. The mapping f is extremal, but not uniquely extremal. This proves the second part of theorem 1.

11. To prove unique extremality in the general case, let ϕ be a rational quadratic differential on a compact Riemann surface R, with at least one second order and no higher order poles. Assume that the trajectories around the second order poles are closed. Let E be a denumerable set of points on R which contains the poles of ϕ and such that the set of accumulation points of E is equal to the set of second order poles of ϕ. The local mappings $z \to \zeta^*$

$$\zeta = \xi + i\eta = \Phi(z), \quad \zeta^* = F_K(\zeta) = \xi + i\frac{1}{K}\eta \quad \left(\Phi(z) = \int \sqrt{\phi(z)}dz\right) \quad (32)$$

generate a new Riemann surface R' and a Teichmüller mapping f of R onto R', with the complex dilatation $\kappa = k\frac{\bar{\phi}}{|\phi|}$. The associated quadratic differential ψ on R' is determined by the local conformal mappings Ψ of R', $\psi = \Psi'^2$, and the local representation of f is

$$f = \Psi^{-1} \circ F_K \circ \Phi. \tag{33}$$

The set E is mapped onto a set $E' = f(E)$ on R'.

Let $\tilde{f} : R \to R'$ be qc, $\tilde{f} = f$ on E and $\tilde{f}$ homotopic to f in $R \setminus E$. Moreover, assume that $\tilde{f}$ is extremal, hence $\tilde{K} = K$, since by section 5 f is extremal.

The proof that $\tilde{f} = f$ by the length area method is based on a partition of R into ϕ-rectangles and ϕ-annuli. Let G_n be a subdomain of R bounded by closed trajectories lying in the neighborhoods of the second order poles. For any closed trajectory α_j of ϕ in G_n, of length a_j, the image curve by $\tilde{f}$ has ψ-length at least a_j (the image of α_j by f is a closed trajectory of ψ of length a_j.) We get

$$a_j \leq \int_{\alpha_j} |d\zeta^*| = \int_{\alpha_j} |\Psi'(w)| \, |p\, dz + q\, d\bar{z}|, \tag{34}$$

with $dw = p\, dz + q\, d\bar{z}$, $w = \tilde{f}(z)$. z and w are, at this stage, arbitrary local parameters on R and R' respectively. Introducing the parameter $\zeta = \Phi(z)$ along α_j we get

$$a_j \leq \int_{\alpha_j} \frac{|\Psi'(w)|}{|\Phi'(z)|} \, |p + q\frac{\phi(z)}{|\phi(z)|}| \, d\xi \tag{35}$$

and hence

$$\left.\begin{aligned}
|R_j|_\phi = a_j b_j &\leq \iint\limits_{R_j} \frac{|\Psi'(w)|}{|\Phi'(z)|} |\, p + q\frac{\phi(z)}{|\phi(z)|} |\, d\xi\, d\eta \\
&= \iint\limits_{R_j} |\psi(w)|^{\frac{1}{2}} |\phi(z)|^{\frac{1}{2}} |p + q\frac{\phi(z)}{|\phi(z)|}|dx\, dy.
\end{aligned}\right\} \tag{36}$$

Here, R_j is the annulus swept out by the closed trajectories homotopic to α_j, b_j is its ϕ-height and hence $|R_j|_\phi = a_j b_j$ its ϕ-area.

For a spiral set S_K (limit set of a recurrent trajectory ray) we choose a vertical interval β_k and cover S_k by the horizontal rectangles $S_{k\nu}$ based on β_k (for details, see [6], pg 52). Since any two points of the f-image of β_k have ψ-distance smaller than any positive number ε, if β_k is chosen sufficiently short, the length inequality is approximately valid. For a horizontal interval $\alpha_{k\nu}$ in $S_{k\nu}$ with ϕ-length $a_{k\nu}$ we find

$$a_{k\nu} - \varepsilon \leq \int\limits_{\alpha_{k\nu}} |d\zeta^*| = \int\limits_{\alpha_{k\nu}} |\Psi'(w)| \, |p\, dz + q\, d\bar{z}|. \tag{37}$$

Integration over $S_{k\nu}$ gives the inequality (36) in the form

$$|S_{k\nu}|_\phi - \varepsilon b_{k\nu} \leq \iint\limits_{S_{k\nu}} |\psi(w)|^{\frac{1}{2}} |\phi(z)|^{\frac{1}{2}} |p + q\frac{\phi(z)}{|\phi(z)|}|dx\, dy, \tag{38}$$

where $b_{k\nu}$ is the ϕ-height of the rectangle $S_{k\nu}$. Summation over all ν leads to

$$|S_k|_\phi - \varepsilon \cdot b_k \leq \iint\limits_{S_k} |\psi(w)|^{\frac{1}{2}} |\phi(z)|^{\frac{1}{2}} |p + q\frac{\phi(z)}{|\phi(z)|}|dx\, dy, \tag{39}$$

with b_k the ϕ-length of β_k. Another summation over all the finitely many annuli and spiral sets of G_n gives

$$|G_n|_\phi - \varepsilon \cdot b \leq \iint\limits_{G_n} |\psi(w)|^{\frac{1}{2}} |\phi(z)|^{\frac{1}{2}} |p + q\frac{\phi(z)}{|\phi(z)|}|dx\, dy, \tag{40}$$

where $b = \sum b_k$. Since the β_k can be chosen arbitrarily short, ε can be made arbitrarily small. This affects of course the decomposition of the S_k into horizontal rectangles, but it does not affect the right hand integral, although it was originally obtained by means of a strip decomposition. Therefore

$$|G_n|_\phi \leq \iint\limits_{G_n} |\psi(w)|^{\frac{1}{2}} |\phi(z)|^{\frac{1}{2}} |p + q\frac{\phi(z)}{|\phi(z)|}|dx\, dy. \tag{41}$$

We apply the Schwarz inequality to the right hand side, after multiplication above and below by the square root of the Jacobian $J(w/z) = |p|^2 - |q|^2$ of $\tilde{f}$. We get

$$|G_n|_\phi^2 \leq \iint\limits_{G_n} |\psi(w)| J(w/z) dx\, dy \cdot \iint\limits_{G_n} |\phi(z)| \frac{|p + q\frac{\phi(z)}{|\phi(z)|}|^2}{|p|^2 - |q|^2} dx\, dy. \qquad (42)$$

The first factor is the ψ-area of $\tilde{f}(G_n)$. The ψ-area of $f(G_n)$ is $\frac{1}{K}|G_n|_\phi$, as is readily seen from the local representation of the mapping f and the fact that the ψ-area is the Euclidean area in the ζ^*-plane. According to section 9, the ψ-area of $\tilde{f}(G_n)$ satisfies

$$|\tilde{f}(G_n)|_\psi \leq \frac{1}{K}|G_n|_\phi + \varepsilon, \qquad (43)$$

with ε arbitrarily small with $G_n \to R$.

The integrand of the second factor is

$$\left.\begin{aligned}
|\phi|\frac{|1 + \kappa\frac{\phi}{|\phi|}|^2}{1 - |\kappa|^2} &= |\phi|\frac{1 + |\kappa|}{1 - |\kappa|} - 2\frac{|\kappa|\,|\phi| - \mathrm{Re}\,\kappa\phi}{1 - |\kappa|^2} \\
&\leq |\phi| \cdot K - \frac{2}{1 - k^2} \cdot \{k|\phi| - \mathrm{Re}\,\kappa\phi\}.
\end{aligned}\right\} \qquad (44)$$

The inequality (42) now becomes

$$|G_n|_\phi^2 \leq \left\{\frac{1}{K}|G_n|_\phi + \varepsilon\right\}\left\{K|G_n|_\phi - \frac{2}{1 - k^2}\iint\limits_{G_n}(k|\phi| - \mathrm{Re}\,\kappa\phi)dx\, dy\right\}. \qquad (45)$$

After subtraction of the term $|G_n|_\phi^2$ we get

$$0 \leq \varepsilon K|G_n|_\phi - \frac{1}{K}|G_n|_\phi\frac{2}{1 - k^2}\iint\limits_{G_n}(k|\phi| - \mathrm{Re}\,\kappa\phi)dx\, dy. \qquad (46)$$

Dividing by $|G_n|_\phi$, we see that

$$\iint\limits_{G_n}(k|\phi| - \mathrm{Re}\,\kappa\phi)dx\, dy \leq \varepsilon \cdot \frac{K^2(1 - k^2)}{2}. \qquad (47)$$

This is true for all sufficiently large n, and since the integrand is non negative, we get

$$\iint\limits_{R}(k|\phi| - \mathrm{Re}\,\kappa\phi)dx\, dy \leq \varepsilon, \qquad (48)$$

which is in term true for all $\varepsilon > 0$. We conclude that

$$\iint\limits_{R} (k|\phi| - \mathrm{Re}\,\kappa\phi)dx\,dy = 0, \qquad (49)$$

hence $\kappa\phi = k|\phi|$ and thus

$$\kappa = k\frac{\overline{\phi}}{|\phi|}. \qquad (50)$$

The mappings f and $\tilde{f}$ have the same complex dilatation and therefore coincide up to a conformal mapping, which necessarily is the identity.

If there exists an isolated second order pole P_ν but not all the second order poles are isolated, f is extremal but not uniquely extremal, since it can be replaced by a conformal mapping near P_ν. The construction is the same as in the plane, using the distinguished parameter. This proves the second part of theorem 2.

Remark. Theorem 2 was stated for quadratic differentials with second order poles all of whose leading coefficients are negative, i.e. the trajectories of ϕ and hence also of ψ near the second order poles are closed.

If the leading coefficients are all positive, we can consider the inverse mapping f^{-1}. The maximal stretching effectuated by f^{-1} is along the orthogonal trajectories of ψ; we are therefore back to the first case.

Assume now that ϕ has positive *and* negative leading coefficients of its second order poles. Again, f is extremal if and only if at least one of the second order poles is an accumulation point of E. Here, the proof is the same, since we have concentrated on one second order pole: we can, if necessary, pass to f^{-1}.

To prove unique extremality if all second order poles are accumulation points of E we now must stick to ϕ. Assume that $\tilde{f}$ is extremal, i.e. $\tilde{K} = K$. Near a second order pole with negative coefficient, the closed trajectories α of ϕ are mapped by $\tilde{f}$ onto curves which are arbitrarily close to the closed trajectories $f(\alpha)$ of ψ. The same is now true for $\tilde{f}^{-1}$ for the closed orthogonal trajectories of ψ: they differ arbitrarily little from their f^{-1} images. But then, it is easy to see that the closed orthogonal trajectories β of ϕ also are mapped, by $\tilde{f}$, onto curves arbitrarily close to $f(\beta)$. The reasoning is performed in terms of the distinguished parameter for ϕ and ψ (it is the same!).

The domains $G_n \subset R$ result from cutting holes around the second order poles, either along closed trajectories or closed orthogonal trajectories. The trajectory structure of ϕ can now also contain cross cuts, running from one closed orthogonal trajectory to another one, possibly the same (for details see [6]). The length-area method applies in just the same way, taking into consideration also the strips of cross cuts. Thus, theorem 2 holds in this more general form.

An example of a rational quadratic differential with a mixed structure, in the plane, is

$$\phi(z) = \frac{1}{z^2} + \frac{1}{z+1} - \frac{1}{z-1}.$$

It has a second order pole at infinity with leading coefficients -1: this is readily seen using the invariance of $\phi(z)dz^2$.

References

[1] O. Lehto and K. Virtanen: Quasiconformal mappings in the plane, Springer-Verlag 1973, 1–258.

[2] E. Reich and K. Strebel: Extremal quasiconformal mappings with given boundary values, Contributions to Analysis, Edited by L. Ahlfors et al., Academic Press 1974, 375–391.

[3] K. Strebel: Zur Frage der Eindeutigkeit extremaler quasikonformer Abbildungen des Einheitskreises, Comment. Math. Helv. 36 (1962) 306–323.

[4] K. Strebel: Zur Frage der Eindeutigkeit extremaler quasikonformer Abbildungen des Einheitskreises II, Comment. Math. Helv. 39 (1964) 77–89.

[5] K. Strebel: On quasiconformal mappings of open Riemann surfaces, Comment. Math. Helv. 53 (1978) 301–321.

[6] K. Strebel: Quadratic Differentials, Erg. d. Math. u. ihrer Grenzgebiete 3. Folge, Bd. 5, Springer-Verlag 1984, 1–184.

[7] K. Strebel: On the extremality and unique extremality of quasiconformal mappings of a parallel strip, Revue Roumaine de mathématiques pures et appliquées, vol. 32, no 10 (1987).

[8] O. Teichmüller: Untersuchungen über konforme und quasikonforme Abbildung, Deutsche Math. 3 (1938) 621–678.

Mathematisches Institut, Universität Zürich

Complex Analysis	Edited by J. Hersch and A. Huber	Birkhäuser Verlag Basel 1988

Lo Yang

Angular Distribution of Meromorphic Functions in the Unit Disk

§1 Introduction

In 1959, W.K. Hayman [1] obtained a series of interesting results on Picard exceptional values of meromorphic functions. Among others, he proved

Theorem A. *Let $f(z)$ be a transcendental meromorphic function in the finite plane. If k is an integer not less than 5 and a is a finite non-zero complex value, then $f' - af^k$ assumes every finite complex value infinitely often.*

The criterion for normality which corresponds to Theorem A was recently proved by J.K. Langley [2] and Li Xianjin [3] respectively.

For meromorphic functions in the unit disk, it is easy to prove the following proposition by Hayman's method [1].

Theorem B. *Let $f(z)$ be meromorphic and of order λ in $|z| < 1$, where $0 < \lambda < +\infty$. If k is an integer not less than 5 and a ($\neq 0$) and b are two finite complex values, then we have*

$$\limsup_{r \to 1} \frac{\log n(r, f' - f^k = b)}{\log \frac{1}{1-r}} = \lambda + 1. \tag{1.1}$$

In this note, we shall prove such a result in the angular distribution.

Theorem. *If $f(z)$ is given as in Theorem B, then there exists a point $e^{i\theta_0}$ $(0 \le \theta_0 \le 2\pi)$ such that the equality*

$$\limsup_{r \to 1} \frac{\log \bar{n}(r, \theta_0, \varepsilon, f' - af^k = b)}{\log \frac{1}{1-r}} = \lambda + 1 \tag{1.2}$$

holds for any small positive number ε, every positive integer k not less than 5 and two arbitrary finite complex values a $(\ne 0)$ and b, where $\bar{n}(r, \theta_0, \varepsilon, f' - af^k = b)$ denotes the number of zeros of $f'(z) - af(z)^k - b$ in the region $(|z| \le r) \cap (|\arg z - \theta_0| \le \varepsilon)$, multiple zeros being counted only once.

§2   Two Lemmas

Lemma 1. *Suppose that $f(z)$, $a(z)$ and $b(z)$ are meromorphic functions in $|z| < R$ $(0 < R \le +\infty)$ and that k is an integer not less than 5. If*

$$\Psi(z) = \frac{f'(z) - b(z)}{a(z)f(z)^k}$$

and $\Psi(0) \ne 0, \infty,$[1]$)$ then we have

$$\left.\begin{aligned}
(n-2)T(r,f) &\le T(r, \Psi) - N_0(r) + m\left(r, \frac{f'}{f}\right) \\
&\quad + T\left(r, \frac{1}{a}\right) + T(r, b) + \log \frac{1}{|\Psi(0)|} + \log 2
\end{aligned}\right\} \tag{2.1}$$

for $0 < r < R$, where $N_0(r)$ is the counting function corresponding to $n_0(r)$ and $n_0(r)$ denotes the number of zeros of $f'(z) - b(z)$ in $|z| \le r$ which are not zeros of $f(z)$.

Lemma 2. *With the same assumptions and notations as Lemma 1, if $\Psi(0) \ne 1$, $\Psi'(0) \ne 0$ and $f'(0) - b(0) \ne 0, \infty$, then we have*

$$(n-4)T(r,f) \le \bar{N}\left(r, \frac{1}{\Psi - 1}\right) + S(r, f, \Psi), \tag{2.2}$$

---

[1]) The assumptions on the initial value $\Psi(0)$ are not essential. Otherwise, it can be replaced by the first non-zero coefficient of the Taylor expansion of $\Psi(z)$ at the origin.

where

$$
\left.
\begin{aligned}
S(r, f, \Psi) = {} & T\left(r, \frac{1}{a}\right) + T(r, b) + k(N(r, a) \\
& + N\left(r, \frac{1}{a}\right) + N(r, b)) + 2m\left(r, \frac{f'}{f}\right) \\
& + m\left(r, \frac{\Psi'}{\Psi}\right) + m\left(r, \frac{\Psi'}{\Psi - 1}\right) + \frac{1}{k}\log\frac{1}{|f'(0) - b(0)|} \\
& + \frac{1}{k}\log|\Psi(0)| + \log\left|\frac{\Psi(0) - 1}{\Psi'(0)}\right| + 2\log 2.
\end{aligned}
\right\}
\tag{2.3}
$$

The proofs of Lemma 1 and Lemma 2 can be carried through as Hayman [1] and Yang [6].

§3 Proof of the Theorem

According to the Nevanlinna second fundamental theorem, we have

$$
\limsup_{r \to 1} \frac{\log N(r, f = \alpha)}{\log \frac{1}{1-r}} = \lambda
\tag{3.1}
$$

for any complex value α with the exception of two values at most. Let α_0 be a fixed non-exceptional value. It follows from (3.1) and a known fact [4, 263–264] that

$$
\limsup_{r \to 1} \frac{\log n(r, f = \alpha_0)}{\log \frac{1}{1-r}} = \lambda + 1.
$$

Thus there exists a number θ_0 $(0 \le \theta_0 < 2\pi)$ such that the equality

$$
\limsup_{r \to 1} \frac{\log n(r, \theta_0, \delta, f = \alpha_0)}{\log \frac{1}{1-r}} = \lambda + 1
\tag{3.2}
$$

holds for any positive number δ. We shall prove that the point $e^{i\theta_0}$ satisfies relation (1.2) of the Theorem.

In fact, we suppose first that $\theta_0 = 0$. The general situation can be reduced to this case by a suitable rotation.

Following G. Valiron [5] [7], we consider two sectors

$$
S : (|z| < 1, \, |\arg z| < \varepsilon)
$$

and

$$S_M : \left(|z| < 1,\ |\arg z| < \frac{\varepsilon}{M}\right),$$

where M is a sufficiently large positive number.

The transformation

$$z_1 = z_1(z) = z^{\pi/2\varepsilon} \tag{3.3}$$

maps S onto the right half-disk D: $(|z_1| < 1, |\arg z_1| < \frac{\pi}{2})$ and S_M onto D_M: $(|z_1| < 1, |\arg z_1| < \frac{\pi}{2M})$. Set $z(z_1) = z_1^{2\varepsilon/\pi}$ and $f_1(z_1) = f(z(z_1))$.

Let z_1 be the image of z under the mapping (3.3). If the moduli of z and z_1 are r and r_1 respectively, then we have

$$1 - r = 1 - r_1^{2\varepsilon/\pi} = \frac{2\varepsilon}{\pi}(1 - r_1)(1 + O(1 - r_1)). \tag{3.4}$$

Since the equality (3.2) holds for any positive number δ and $\theta_0 = 0$, we obtain

$$\limsup_{r_1 \to 1} \frac{\log n(r_1, 0, \frac{\pi}{2M}, f_1 = \alpha_0)}{\log \frac{1}{1-r_1}} = \lambda + 1. \tag{3.5}$$

Further, the transformation

$$\frac{z_2 - i}{z_2 + i} = i\left(\frac{z_1 - i}{z_1 + i}\right)^2 \tag{3.6}$$

maps the right half-disk D to $|z_2| < 1$. (3.6) can be written as

$$z_2 = z_2(z_1) = -\frac{z_1^2 + 2z_1 - 1}{z_1^2 - 2z_1 - 1}. \tag{3.7}$$

Denoting by $z_1 = z_1(z_2)$ the inverse mapping of (3.6), we set

$$f_2(z_2) = f_1(z_1(z_2)) = f(z(z_1(z_2))).$$

It follows from (3.7) that

$$1 - |z_2|^2 = \frac{8\mathrm{Re}(z_1)(1 - |z_1|^2)}{|z_1|^4 + 4|z_1|^2 + 1 + 4\mathrm{Re}(z_1)(1 - |z_1|^2) - (z_1^2 + \bar{z}_1^2)}.$$

When z_1 tends to 1, $1 - |z_2|$ is very close to $2(1 - |z_1|)$. Thus (3.5) gives

$$\limsup_{r_2 \to 1} \frac{\log n(r_2, f_2 = \alpha_0)}{\log \frac{1}{1-r_2}} \geq \lambda + 1. \tag{3.8}$$

By a known result [4, 263–264], we have

$$\limsup_{r_2 \to 1} \frac{\log N(r_2, f_2 = \alpha_0)}{\log \frac{1}{1-r_2}} \geq \lambda.$$

Therefore, the order of f_2 in $|z_2| < 1$ is not less than λ.

Using (2.2) and (2.3) of Lemma 2 to $f_2(z_2)$ in $|z_2| < 1$, we obtain

$$\limsup_{r_2 \to 1} \frac{\log \bar{N}(r_2, f_2' - a(z_2)f_2^k = b(z_2))}{\log \frac{1}{1-r_2}} \geq \lambda, \tag{3.9}$$

where $a(z_2)$ and $b(z_2)$ are two arbitrary meromorphic functions of order less than λ and k is any integer not less than 5.

Let $z_2 = z_2(z)$ be the compound transformation of (3.3) and (3.6) and let $z = z(z_2)$ be its inverse mapping. If a $(\neq 0)$ and b are two finite complex values, we choose

$$a(z_2) = az'(z_2), \qquad b(z_2) = bz'(z_2).$$

Denoting by z_2 the image of z, it is clear that

$$z'(z_2) = \frac{1}{z_2'(z)}.$$

When $z \in S$, then we have $z_1 \in D$ and

$$|z_1'(z)| \leq \frac{\pi}{2\epsilon} \tag{3.10}$$

by (3.3). Noting

$$z_2'(z_1) = \frac{4(z_1^2 + 1)}{(z_1^2 - 2z_1 - 1)^2}$$

from (3.7), we deduce that

$$|z_2'(e^{i\theta})| \leq \frac{8}{4 + 2(1 - \cos 2\theta)} \leq 2, \quad (|\theta| \leq \frac{\pi}{2}),$$

and

$$|z_2'(ci)| = \frac{4(1 - c^2)}{(1 + c^2)^2 + 4c^2} \leq 4, \quad (|c| \leq 1).$$

Thus

$$|z_2'(z_1(z))| \leq 4, \qquad (z \in S). \tag{3.11}$$

Comparing (3.10) and (3.11), we have

$$|z_2'(z)| = |z_2'(z_1(z)) \cdot z_1'(z)| \leq \frac{2\pi}{\varepsilon},$$

so that

$$T(r_2, z'(z_2)) = T\left(r_2, \frac{1}{z_2'(z(z_2))}\right)$$

$$= T(r_2, z_2'(z(z_2))) + O(1)$$

$$\leq \log \frac{2\pi}{\varepsilon} + O(1).$$

Therefore the function $z'(z_2)$ is of order zero in $|z_2| < 1$. It follows from (3.9) that

$$\limsup_{r_2 \to 1} \frac{\log \bar{n}(r_2, f_2' - az'(z_2)f_2^k = bz'(z_2))}{\log \frac{1}{1-r_2}} \geq \lambda + 1. \qquad (3.12)$$

Since

$$1 - r_2 = \left|e^{i \arg z_2} - z_2(z_1)\right| \leq \left|z_2(e^{i \arg z_1}) - z_2(z_1)\right|$$

$$= \left|\int_{e^{i \arg z_1} z_1} z_2'(z_1) dz_1\right| \leq 4\left|e^{i \arg z_1} - z_1\right| = 4(1 - r_1),$$

we have by (3.12) that

$$\limsup_{r_1 \to 1} \frac{\log \bar{n}\left(r, 0, \frac{\pi}{2}, \frac{f_1'(z_1)}{z_2'(z_1)} - az'(z_2(z_1))f_1(z_1) = bz'(z_2(z_1))\right)}{\log \frac{1}{1-r_1}} \geq \lambda + 1.$$

Noting (3.4), we obtain finally

$$\limsup_{r \to 1} \frac{\log \bar{n}\left(r, 0, \varepsilon, P(z, f) = bz'(z_2(z_1(z)))\right)}{\log \frac{1}{1-r}} \geq \lambda + 1,$$

where

$$P(z, f) \equiv \frac{f'(z)}{z_2'(z_1(z))z_1'(z)} - az'(z_2(z_1(z)))f(z)^k.$$

Therefore

$$\limsup_{r \to 1} \frac{\log \bar{n}(r, 0, \varepsilon, f' - af^k = b)}{\log \frac{1}{1-r}} \geq \lambda + 1. \qquad (3.13)$$

But the order of $f' - af^k$ is λ, the left-hand side of (3.13) does not exceed $\lambda + 1$. The proof of the Theorem is complete. $\square$

References

[1] Hayman W.K., Picard values of meromorphic functions and their derivatives, *Ann. of Math.*, 70 (1959), 9–42.

[2] Langley J.K., On normal families and a result of Drasin, *Proc. Royal Soc. Edinburgh*, 98A (1984), 385–393.

[3] Li Xianjin, Proof of a conjecture of Hayman, *Sci. Sinica*, Series A, 28 (1985), 596–603.

[4] Nevanlinna R., *Analytic functions*, Springer-Verlag, New York, 1970.

[5] Valiron G., Points de Picard et points de Borel des fonctions méromorphes dans un cercle, *Bull. Sci. Math.*, 2e série 56 (1932), 10–32.

[6] Yang Lo, Normal families and differential polynomials, *Sci. Sinica*, Series A, 26 (1983), 673–686.

[7] Yang Lo et Shiao Shiou-zhi, Sur les points de Borel des fonctions méromorphes et de leurs dérivées, *Sci. Sinica*, 14 (1965), 1556–1573.

Institute of Mathematics
Academia Sinica
Beijing, China